Envisioning TeleCity

Frank Helten/Bernd Fischer (eds.)

ENVISIONING TELECITY

Towards the Urbanisation of ICT

PETER LANG
Frankfurt am Main · Berlin · Bern · Bruxelles · New York · Oxford · Wien

Bibliographic Information published by Die Deutsche Bibliothek
Die Deutsche Bibliothek lists this publication in the Deutsche Nationalbibliografie; detailed bibliographic data is available in the internet at <http://dnb.ddb.de>.

ISBN 3-631-51056-X
US-ISBN 0-8204-6453-8

www.peterlang.de

Contents

Preface

This is a book about visions of the city in the future. What are the possible forms and directions of the interplay of information and communication technologies (ICT) and urban developments? Here we have collected a range of contributions to the ongoing discussion of how the city will change in the information society.

The basis of this publication was the international conference "Envisioning TeleCity – the Urbanisation of ICT" held in December 2001 in Berlin. The papers and discussions, both during and after the conference, were so lively that the participants encouraged us to publish an edited volume under the title of the conference.

Why hold yet another conference on the subject of ICT and cities? In the early 1980s, some scholars anticipated that the rise of the information society would have profound implications for cities. However, the interrelationship between information and communication technologies (ICT) and the city remained largely unclear and invisible. By the 1990s, we watched ICT progress to surveillance tool, while the web had come to serve as a complex resource for interpersonal communication, information, advertising, and the like. The goal of the conference "Envisioning TeleCity" was to develop an innovative approach to issues we have followed since the 1980s. Today's world has changed fundamentally; ICT is omnipresent. The conference aimed to serve as an exchange forum for academics, politicians and planners interested in the interrelation of ICT and cities. It also gave the TeleCityVision research consortium the opportunity to present the main findings of the EU research project "TeleCityVision – Information Society and Urban Development in European Comparison" to a broader public.

The conference also attempted to combine new approaches for the investigation of this field of research, and to discuss findings from various international studies. We at the Berlin Institute for Social Research asked the group of approximately 50 invited scholars and experts from throughout Europe to concentrate on four main points: 1. Changes in urban structures prompted by the increased use of ICT, 2. The role of the public sector and its responsibility as actor in the informatisation of society, 3. The relevance of ICT for economic policy at the communal level, and 4. The role of ICT in urban planning.

The structure of this volume tries to reflect the results of the conference discussions. In Part One, Florian Rötzer, Stephen Graham, Saskia Sassen, Willem van Winden and André van der Meer outline viable bases for articulating visions of the future of ICT, and use recent developments to illustrate how these visions are becoming reality in a range of current projects, policies and strategies.

Florian Rötzer, editor of the German net-journal *Telepolis*, raises the question of how we imagine the interrelation of ICT with the socio-spatial sphere and lifestyles of different actors in the city.

In their paper titled "Planning Cyber-Cities? Integrating Telecommunications into Urban Planning," Stephen Graham and Simon Marvin from the School of Architecture, Planning and Landscape at the University of Newcastle take a systematic view of the relation between telecommunications and the city. They discuss the possible effects of and approaches for "...creatively integrating telecommunications into urban policy and planning practices and strategies."

Saskia Sassen from the University of Chicago turns the spotlight on the embeddedness of digital technologies and investigates the possible structural and political implications of this fact on the city. Sassen's analyses lead her to conclude that the application of digital technologies affects the balance of power and influence, creating new imbalances while simultaneously stabilising old ones. As a possible solution, she presents the concept of a counter-geography, and shows that digital technologies have considerable potential to mobilise counter-publics and counter-cultures.

The Dutch economists André van der Meer and Willem van Winden concentrate on the possibility of economic growth and more employment through digital technologies. In their contribution "The ICT Sector: Growth Engine for Urban Regions?" they present the conditions and development factors in different ICT clusters in three European urban regions. Against this background, they discuss starting points for future concepts.

In Part Two, Bernd Fischer, Galit Cohen-Blankshtain, Peter Nijkamp, Manuel Valenzuela, Carmen Vázquez, Santiago Lorente, Leon Hempel and Hans-Liudger Dienel took the results from the TeleCityVision Project and analysed them from various perspectives. The object of the analysis was to identify the details of the visions governing the policy-makers and public administrators of several European cities as they design and implement ICT-related strategies.

During research, it became increasingly obvious that traditional parameters and dimensions of describing and categorising cities and urban developments no longer suffice to analyse the immaterial and barely visible process which may be characterised as the urbanisation of ICT.

Bernd Fischer, social researcher at the Berlin Institute of Social Research, sees the urbanisation of ICT as an organisational and political learning process, a process that could offer the public actors involved different means of overcoming an ICT induced cultural lag. In his paper "Urban authorities and the harnessing of ICT," he takes empirical findings on a micro-social level, and shows us what various strategies of harnessing can lead to by comparing developments in different city-types.

The Israeli geographer Galit Cohen-Blankshtain and the Dutch economist Peter Nijkamp attempt to measure ICT-related beliefs among urban actors. In their paper titled "Measuring beliefs about ICT: the case of the European front liners," they develop a model to assist in identifying the belief patterns of urban administrators and politicians. The model enables them to found assumptions about whether and to what extent urban actors are prepared to implement adequate ICT policies.

In collaboration with the Spanish sociologist Santiago Lorente, the Spanish geographers Manuel Valenzuela and Carmen Vasquez focus on the paired expressions "network city and city networking." They analyse the intra and interurban dimensions of the process by reconstructing existing socio-spatial patterns of the diffusion and application of ICT in cities and urban systems.

In the final contribution, the German political scientist Leon Hempel and the historian of technology and engineer Hans-Liudger Dienel investigate the perception of ICT under increasing urban competition. They conclude that the potential of ICT in urban developments and processes can only be fully exploited if the visions and assumptions of urban actors regarding ICT applications for different purposes are coupled with an appropriate concept of the post-fordistic city. This concept should balance economic and social necessities and lead to sustainable urban development.

We are well aware that the scope of this volume is limited to presenting a pattern of different approaches and views on the subject. However, we hope all those interested in the interrelation of ICT and urban development will find it a useful basis for discussion.

Acknowledgements

Many people contributed to make the conference happen and this book possible. We would first like to thank the European Commission and Berlikomm for their financial support of the conference. We also received generous support in organizational matters from a number of individuals. Among our colleagues at the Berlin Institute for Social Research we would like to express our gratitude to Gaby Senftleben, Steffi Körner, Anne Röhrig, Gunnar Schwan and Dirk Faber for their excellent help and advice, as well as our directors Sibylle Meyer and Eva Schulze. Many thanks also to Leon Hempel and Hans-Liudger Dienel from the Centre for Technology and Society at the TU-Berlin for organisational support.

It is not, of course, possible to name all the individuals who helped inspire the conference and the TeleCityVision research project. But we would like to single out Sandra Huning, who was part of the German team during the whole empirical phase, and the entire TCV-Consortium, including their patience with us as we gathered experience in co-ordinating research at the European level, for our thanks: Galit Cohen-Blankshtain, Hans-Liudger Dienel, Roderick Flynn, Pascal Gillon, Leon Hempel, Angelika Kofler, Santiago Lorente, Oumar N'Diaye, Peter Nijkamp, Ronald Pohoryles, Ann Rudinow Sætnan, Manuel Valenzuela, and Carmen Vázquez.

We owe a further debt to the many interviewees across Europe for the insights they garnered from their knowledge and visions of urban life, and generously shared with us.

We are also grateful to Liverpool University Press for kind permission to reproduce Stephen Graham's and Simon Marvin's article "Planning

Cybercities? Integrating Telecommunications into Urban Planning" which was first published in Town Planning Review (1) 1999.

Frank Helten and Bernd Fischer

Berlin, May 2003

2 Restructuring Cities

Florian Rötzer

With its infrastructure, gates, and strongholds, cyberspace is establishing itself in the real world at a pace so breathtaking it is hard to comprehend. The trend is clear: as technology becomes ever more miniature, networks can encompass an incredible range of things, from the intelligent house and its array of appliances, such as networked and intelligent refrigerators and garbage cans, to robots and cars, and even animals and people. Microchips, sensors, effectors, transmitters, cables, and satellites are crisscrossing the globe like a second skin, wrapping us in a gigantic flood of circulating data. Over the longer term, everything that has benefited from the advantages of spatial concentration and centralization is in danger of losing its importance. Here we are talking not only about the role of global competition, as rapid transportation means have made the world smaller; rather, it is computer networks that are threatening our conventional spatial hierarchies. Increasingly we will find that the contexts for interaction, which, until now, defined spatial concentration and made it attractive, will pass into cyberspace, either partially or in their entirety. Or cyberspace will radically redesign how we interact. The exodus of trade, businesses and banks, public agencies and offices, schools and universities, libraries and museums, is already underway. The more broadband networks become, that is, the faster they can transmit data, the more powerful the exodus.

Increasingly, people will live in two worlds at the same time. These worlds interlock and permeate each other, but they also call forth new conflicts and profound changes. One possible scenario is that spatial structures like cities will unravel as "concentration and acceleration machines" in countries with wealth and networks. Spatial structures, as we know them, will make way for the dispersed structures of a placeless "digital urbanism" with its points of access to virtual spaces. At the same time, the "old" cities with their spiraling financial deficits will become receptacles for the refuse of society. Digital urbanism means that in the future, most life will take place in decentral metropolitan landscapes, where the conventional distinction between city and country will become increasingly fuzzy – a blurring we already see today.

The circumstances behind my invitation to this conference might themselves indicate how the significance of cities and spaces with a high concentration of people, institutions, and buildings could gradually change. As you know, Saskia Sassen was supposed to have come in my place. The terrorist attacks on the metropolis New York, particularly on the symbols of urbanity in the form of the World Trade Center towers, destroyed not only a piece of urbanity. Airplanes, for their part, our most rapid means of mass transportation, revealed themselves as very risky. Means of mass transportation and concentrated urban spaces, in which masses of people work and live, turned into threats. Ever since the discovery of gunpowder and long-range weapons, cities and their fortifications ceased to offer protection against outside attacks. With air bombardments and the rise of means of mass destruction, cities became traps. Compared with rural areas, cities offered a maximum of destructive possibilities and collected, in most cases, key aspects of important national infrastructure for politics, the military, economy, and society. Since then, the spatially concentrated mass is itself a potential risk and, of course, a genuine target for both terrorists and military attacks. At the same time, however, cities also permit small groups to go underground, since large cities resemble, in this sense, impenetrable places such as jungles or mountain landscapes. Following the attacks with passenger plane bombers, moreover, we were confronted by the first anthrax-infected letters, which showed that each and every letter or package might pose a dangerous missive for the recipient. Finally, another airplane crashed, regardless of the reasons why, in a residential neighborhood of New York. Saskia Sassen then decided, for understandable reasons, not to attend this conference. In light of this recent past, one will, in the future, be inclined to consider whether one absolutely needs to travel to a specific location when it might be possible to communicate by other means, for example by video conference. Often, of course, we still prefer that people are physically present, but we stand only at the threshold of the tele-society epoch, while we have long grown accustomed to talking on the phone or watching other people on screens. The rapid inroads of cell phones, heralds of mobile and ever-present networking, have shown that communicating with individuals far away is often enough more important than communicating with people close to home. At the very least, the permanent ability to connect, regardless of time or place, together with the possibility of mutual surveillance, show that virtuality and real space are pervading each other to an ever greater extent – and this will, of course, have repercussions for how we construct real space.

Perhaps the passenger planes, transformed into bombs, symbolically underscored, as they smashed into the dual landmarks of global economy, the gradual end of skyscrapers and thus our concentrated urban centrality. So, too, could the threats of the anthrax letters, refashioned into deadly missives, allow virtual email to triumph over conventional mail, and encourage the continued virtualization of society – even if this goes hand in hand with such new hazards as cracker attacks, cyber sabotage, or only viruses, sent together with emails. In contrast to physical threats, virtual risks have, until now, remained relatively harmless. We have yet, at least, to experience a devastating cyberterrorist attack. Nonetheless, the US government is already considering the construction of a network completely isolated from the internet, in order to protect important functions. The disadvantage, of course, would be its complete withdrawal from the public sphere and the need to operate two networks parallel to each other. In our media society, it is easy enough to capture attention for certain topics and images – but our media memory is short-lived. Confronted with a terrorist threat, indiscriminate in its choice of possible victims, security measures and defense fantasies proliferate, even though it is likely they will quickly fade as long as there are no new terrorist attacks. We could, however, find that certain developments will gather momentum, longer-term trends such as the shift from spatial centrality to a greater decentrality or, more banal, from letters to emails, from meetings or work contexts in real spaces to those in virtual spaces, or from large buildings to networks.

The collapsed skyscrapers in the middle of Manhattan have, however temporarily, prompted the businesses whose offices were once in one of the buildings to relocate. A total of 1.2 million square meters of office space was annihilated in the 6 buildings destroyed in Manhattan. Today we find the same amount of office space vacant in Lower Manhattan, 50 percent more than before September 11. Most businesses have, of course, relocated within New York City. But what was initially conceived as a temporary solution could well put down roots, especially since real estate costs are lower in other areas and the threat of attacks is smaller. Moreover, it could also become clear that, thanks to information technologies, it is no longer necessary to be present in concentrated urban zones. The assault on the civilized world becomes, in this sense, an assault on the metropolitan cities and thus on spatial centrality. As response to what are thought to be globally dispersed terrorist networks, it is not only the military operations, with distant wars and unmanned fighter- and reconnaissance aircraft, that would increasingly rely on small, covert, decentral and flexible commandos – but also how we choose to live in space. The future might well lie in digital de-urbanization, our large cities constructed, together with their functions, in cyberspace.

Over time, technology will provide ever better means for detaching places from the access to information and communication through digital media. The information city will emigrate, to the extent possible, to virtual space. There it will create a truly global metropolis, anchored throughout the world, accessible wherever one can connect to computer networks. Even today, this virtual metropolis is far larger than the largest city. Likely estimates reach four hundred million inhabitants – most public agencies, businesses, organizations, as well as many individuals have already opened their subsidiaries, at the very least, and have thus relocated. The virtual metropolis relies on the infrastructure, culture, and life of our cities to grow, even as it leaves these behind and absorbs them, just as once our cities spread from their original settlements to the surrounding villages. But here is the decisive turn. For all that cities, through their centrality and concentration, offered all the advantages of increasing speed of traffic and communication, precisely this has gradually shown itself to be a brake, if not a danger to their continued development. Whereas mass transportation had to rely on the network of geographically dispersed and concentrated centers, the transition from the industrial to information society – and information production, processing, and dispatch – has severed our dependence on spatial transport.

Our greater urban areas will not, of course, disappear. Rather, how our society lives (and we ourselves in it) will not necessarily be tied to spatial concentrations. The image, at least, of the city as space with an economic center in the middle continues to dominate us. Nor can we forget that "global cities" have a competitive advantage, as Saskia Sassen has repeatedly emphasized: their infrastructure, for example in the form of rapid data transfer, along with the spatial proximity of the necessary specialists and institutions, as well as other enterprises. "Living" cities are also attractive for what they offer in where we live, play, relax, socialize. The new centrality, via the concentration of telecommunications infrastructure, also allows for the global expansion of business activities, since centrality makes it easy to monitor and direct what businesses do. The advantage that cities, and here the centers, enjoy in terms of infrastructure is probably, however, of limited duration. For some time now, rapid means of transportation and telephones, radios, and television have served to level the differences between city and country. There is no longer an inherent advantage to being present in a city center, as opposed to proximity to good highways, an airport, or train connections.

Simply because of population growth, greater urban areas will continue to represent the space where the vast majority of people in the world live. Many cities not only face continuous growth, of course; particularly cities in developing countries must deal with growth so explosive that it takes us into unprecedented sizes, exploding the boundaries and capacities of traditional cities. Cities continue to expand in a process we call sprawling: a seemingly unordered (at least on the surface) spread beyond what were, at least in Europe, clearly defined city limits into the surrounding areas. What results is something we can call urban regions. Often these regions are still linked to the old cities, which are, to a certain extent, the catalyzing cores of this development. Over time, however, urban regions detach themselves from the old cities, gain their own economic power, often come to outstrip the old cities in terms of their financial power, and give rise to a mixture of urban, village, and rural functions, settlements, and lifestyles. We are still at loss for a precise definition of this phenomenon. Agglomeration, urban regions, metropolitan areas – such terms reflect our search for new categories. It is said that up to 90 percent of people in developed industrial countries are already living in cities, but we often forget that the overwhelming majority of urban dwellers live in such suburbs, even though many still work in the downtown areas.

Perhaps it is an exaggeration to say that we are currently experiencing a ruralization of the city form, or an urbanization of the country. Urbanity as lifestyle is difficult to find in such new population centers, in the betwixt-and-between entities that accommodate the plethora of villages, towns, provincial cities, and small cities. Nor do we find much left of rural lifestyles and communities, even though more and more new housing areas follow the designs of the garden city. Increasingly we see such relatively autonomous units with a traditional center for smaller-scale and, above all, homogenous communities in their ethnic or class composition. In these large urban areas, centers and sub-centers can no longer hold their own as far as independence and autonomy are concerned: although their character still owes a great deal to a spatial access to the transportation networks emanating from cities, more and more it is determined by the flood of information and goods in our mobile information society.

What if possibilities for human interaction such as tele-work, tele training, tele shopping, tele-medicine, or the virtualization of public agencies and institutions come to thrive on a large scale? This shift might well precipitate further social fragmentation and growing homogenization of groups or classes. We will be less compelled to expose ourselves to certain social situations – where we need to get along with colleagues at the office, to encounter others as we run errands or stop at the post office, or even to live in any given place because of where we work or study. Instead, our options will increase for how we can live without being forced to depend on community bonds. And where we can choose, rather than are compelled to live together, the stronger will become the trend that people with similar interests and of similar class or ethnic backgrounds will bond together. In the end, our virtual communities are only a reflection of ghettos and gated communities, where one can decide who comes and goes. We can, at any rate, see that as the pressures to live in conventional communities lessen, the function of the state is undermined. The state must appeal to a common interest of all the citizens within its boundaries, across class differences, beyond class differences, to maintain domestic peace. One possibility to restore unity, at least for a time, was – and still is – a threat or a war.

Surrounded by computers, themselves linked to networks, we can participate – as workers, citizens, consumers – in a previously unimaginable global community; at the same time, we are, in principle, ever easier to monitor, regardless of time or place. This tendency had started well before September 11, but the attacks have lent it new strength. The more aware the machines, the more interactive and networked, the more sensors and effectors integrated within them, the greater the number of traces of our activities and, perhaps, emotions, that such machines can find and evaluate. The cell phone allows us to communicate from (almost) anywhere, or to go online, but it also registers our movements and can even serve as a listening device. Security cameras are spreading and becoming ever more "intelligent." They can identify not only individuals, but also behavior patterns; we also see the rise of biometric ID procedures. Satellites circle the skies above us, able to produce highly detailed images. Secret services eavesdrop, as with Echelon, every communication, throughout the world, that goes through satellite. When we work on networked computer systems, software can monitor and record our behavior at the computer down to the smallest detail: which keys we push, how quickly we do so, which programs we open, to whom we send emails, what we write, where we surf. In addition, all the intelligent environments, which are designed to respond to us, monitor us. Here the range is broad: from developments that allow computers in the context of affective computing to respond not only to the movements of an identified individual, but also to physiological data, via robots or virtual agents who are supposed to communicate with us, all the way to intelligent homes.

The upheavals we find in what used to be separate domains of public and private space will have repercussions for urban spaces as well. Even today, through the media, we are actively participating, from previously private spaces, in a public which is immeasurably larger than that offered by the largest metropolis. Not only in television, but also in our homes and at work, we are exposing ourselves to public scrutiny by installing web cams. Likewise, the anonymity of urban public spaces is dwindling as, for example, video surveillance increases. By way of conclusion, I would like to mention one example that might lead us to discuss what life in information environments could look in the future, whether in the city or in the country.

The Aware Home – Life Under Observation

We cannot, of course, predict how the home of the future will look. But it seems likely that at least the homes of those with the necessary money could look a bit like a Big Brother container – even though such homes must, of course, be informed of their inhabitants' well being (via their every step and jump) in order to respond appropriately to their needs. The prospective pioneers of the networked, intelligent, and personalized homes might surprise us: old, well-to-do people may quite possibly be the first owners. That, at least, is the opinion of the founders of the Aware Home prototype, crammed with sensors and computers.

Scientists at the Georgia Institute of Technology are in the process of building a model home for the future, one that can see, hear, and perceive its inhabitants via contact with a range of sensor technologies such as video, audio, movement, and pressure sensors. In the eyes of the users, the computers are supposed to blend into domestic spaces and objects; the entire home becomes an intelligent environment. The "Aware Home" prototype is a two-story house with four rooms. Initial projections tally at least 60 computers, 25 video cameras, the same number of microphones, 40 pressure sensors, four DSL connections, and one ISDN connection with 2 gigabits/s broadband and a cordless internal network with 11 megabits/s.

With the „perceptive" home, one also, of course, wants to lay foundations for the future and go far beyond conventional intelligent homes, where the home computer merely turns lights off and on or regulates the heating system. The final goal is that the home knows precisely who is there, what they are doing, what they would like to have, and how they would like to have it. At the same time, the users are not supposed to be overwhelmed with a new flood of data; nor should they have the sense that their private sphere has been invaded – what might, however, be difficult to reconcile with the project's goal, which aims at an all-around monitoring on the greatest number of levels possible.

One example of the "Aware Home" technology is the "intelligent floor," equipped with pressure sensors. These sensors can determine when and where someone walks, as well as who it is, since individuals can be identified on the basis of their movements. The home should be able to identify and interpret facial expressions and gestures. Should it prove possible to learn the preferences of the inhabitants, and to create a model of their behavior via observation, then the home can be delivered with all the desired furnishings. For now, the scientists do not seem to be able to move much beyond the point of saying that people might want to sleep in on the weekends – and that the automatic coffee maker should turn on a bit later.

Along with the test persons, here scientists or students, the potential inhabitants of such homes are not necessarily technophile nerds, but rather older people. Nor is this premise so improbable, when one considers our aging society: there are greater numbers of old people, who are living longer and who perhaps also have the money to afford a home in which they can live on their own for as long as possible. It could be that the technological equipment for the "Aware House" is less expensive than care-providers; in any event, the willingness to take care of one's parent or parents in the home, even when there still are children around, is declining. The Japanese are tending more towards a somewhat less expensive robot for remote observation, since the house may well be suitable for monitoring, even if it is far from equipped to handle actual care.

In the project "Aging in Place," the home is designed to recognize possible crises and to place an automatic call for help. The home should also prompt the memory of aging persons or to give relatives the security of knowing that everything is fine and that they do not need to worry about the person who is under the home's care. There are already plenty of ideas in this direction. For example, the scientists have developed a scenario for the forgetful older folks, who might be bustling around the kitchen when the doorbell rings: the home takes a couple of pictures to remind them of what they had been doing when they were interrupted.

"Digital family portraits" are one way of monitoring behavior – and soothing anxious relatives. These portraits represent each family member on a flat monitor: "age appropriate, attractive icons," for instance in the form of butterflies – which age are they meant for? -- are placed inside a frame, and are supposed to give a sense of what the senior citizen is in the middle of doing. Is he spending time with others, or is he alone much of the day? The icons do not only portray a single day, but can also summarize the last several days. The color of the icons reflects activity, social behavior, or health, changes for the better or worse. The "image" can read the status of the inhabitant, every day, or several times a day. The decision not to use web cams seems to be an attempt to keep such monitoring from becoming all too invasive.

It is clear that the goal of this project is to incorporate as many functions as possible into "Aware House," without letting the monitoring itself become too obvious. This example of an intelligent environment can, however, also serve as backdrop for the question as to how close such an environment can get to us, without becoming simply burdensome or making us feel paranoid. It does not take millions of viewers; the feeling of never being completely alone and unobserved at home could itself be too much, even if the computers do not act like virtual persons.

3 Planning Cyber-Cities? Integrating Telecommunications into Urban Planning

Stephen Graham, Simon Marvin

Abstract

This paper attempts to how urban planning strategies can creatively address the complex interlinkages between telecommunications, urban forms and transportation. It has three parts. In the first, we set the context for local, planned intervention on IT and telecommunications by exploring the broad relationships between the development and spread of new information technologies, and the changing form and development of cities, and systems of physical mobility and location. We do this, first, by setting the context by looking in turn at the four key areas of city-telecommunications relations. These are: transport-telecommunication relations; the broad links between urban economies and the so-called 'information economy' ; the relations between urban and so-called 'cyber' cultures ; and the ways in which urban communities and 'virtual,' IT-based, communities are interdependently linked. In part two we go on to review a broad, international range of emerging urban planning and policy initiatives which are trying to shape the articulation between urban built forms and electronic interactions. In North America, Europe, and Asia, three broad types of policy are identified: integrated transport and telecommunications strategies, city-level new media policies, and so-called 'information district' and 'urban televillage' plans. Finally, we conclude by assessing the significance of these policies and suggest ways for creatively integrating telecommunications into urban policy and planning practices and strategies.

Introduction: Urban Planning and the Technological Future of Cities

It is now widely argued that the increasingly pervasive applications of linked computer, media and telecommunications technologies constitute nothing less than a wholesale shift of our economy, society and culture. Social scientists regularly now talk of a new, emerging 'digital age', an 'information society,' or a 'network society' (see Gosling, 1997; Castells, 1996). Such a transition is widely believed to be a new industrial revolution, a societal technological and economic shift across capitalist civilisation of similar magnitude to the industrial revolution through which every aspect of society is transformed (Graham and Marvin, 1996).

As part of this shift, cities, and the corridors between them, are being permeated with widening arrays of telecommunications grids - conventional 'phone networks, wireless and radio systems, cable networks, satellite systems, Internet, data and video networks. These silently and (usually) invisibly underpin booming flows of voice, data, video and images across all walks of city life and development. Indeed, every aspect of the life of advanced industrial cities is now cross-cut with all manner of computerised and 'tele-mediated' communications exchanges, and transactions, most of which are now based on digital principles (meaning that they are based on the streams of zeros and ones used in computers). As Geoff Mulgan (1991) once put it, "the re-definition of the city as a system for producing and switching information is highly visible".

IT and telecommunications networks are thus becoming, in a very real sense, the very sinews of our society. For our profoundly urban societies, and for the whole gamut of actors, agencies, policy makers, organisations, and individuals who currently depend on cities in various ways, the relationship between new media and telecommunications technologies and the future of our cities is clearly of critical importance. But what does this so-called 'digital age' really mean for our cities? What future is there for our urban areas, as more and more of the traditional roles and functions, which first generated the need for urban concentration, seem likely to be possible across distance via advanced telecommunications links? Will our cities face some electronic requiem, some nightmarish *Blade-Runner-* style future of decay and polarisation ? Or can they be power-houses of economic, social and cultural innovation in the new electronic media ? And, perhaps most importantly from the perspective of urban planning, what roles are there for urban policies, plans and strategies, and for urban design, community development and transport policies, within the shift to a so-called 'tele-mediated' urban life, based more and more on all types of on-line interactions?

The co-evolution of cities and electronic interactions is increasingly emphasised in technological debates within academia. Debates about 'cyberspace', telecommunications and the future of cities are currently proliferating within disciplines as diverse as architecture, cultural studies, communications studies, science and technology studies, and urban sociology and geography (see, for example, Mitchell, 1995; OTA, 1995; Graham and Marvin, 1996; Castells, 1996). Attention is increasingly being directed to exploring how the economic, social and cultural aspects of cities interact with the proliferation of advanced computer-based telecommunications networks in all walks of urban life (Graham, 1998). Here, the common 1980s assumption that the new communicational capabilities of new media technologies will somehow 'dissolve' the city has waned. Rather, it is now clear that most IT applications are largely metropolitan phenomena. They are developing out of the older urban regions and are associated with new degrees of complexity within cities and urban systems, as urban areas across the world become relationally combined into a globally-interconnected, planetary metropolitan systems (Graham and Marvin, 1996). Research here now centres on the degree to which city economies can be maintained in a world of on-line electronic flows; the ways in which place-based and 'virtual communities' interact; and the related interactions between urban cultures rooted in traditional public spaces, and 'cybercultures' operating within the virtual spaces accessed from computers (see Mitchell, 1995 and Graham and Marvin, 1996, for reviews).

Despite the central importance of the 'urban' in cyberspace debates, however, issues of urban policy and planning have been virtually absent within both the popular and academic sides of the discussions.

New information and communications technologies are usually seen to be some disembodied, external 'wave' of change. Their urban 'impacts' are usually seen to follow inevitably from their effective "collapse of distance" as a constraint on human life (The Economist, 1995). Such scenarios also usually imply that all cities (say, for Europe, London, Leeds, Charleroi, and Athens) will somehow all be 'impacted' in the same ways. Such so-called 'technological determinism' is attractive because it creates powerful scenarios, clear stories, and because it accords with the dominant experience in the West, where, as Stephen Hill (1988) puts it, the pervasive experience of "technology is one of apparent inevitability".

Questions of agency and local policy and planning therefore tend to be ignored in the simple recourse to either generalised, future-oriented debates, or to macro-level, binary models of societal transformation. In these, new technologies are seen to be somehow autonomously transforming society *en masse* into some new 'information age', 'information society' or 'cyberculture'. As the American geographer Robert Warren (1989; 344) argues, "benign projections [about telecommunications and the future of cities] give little indication that there are significant policy issues which should be on the public agenda". In the emphasis on private entrepreneurship and the transcendence of place, the discourses of the 'information society' thus tend to imply that local municipalities, policy agencies and planners might even be little more than irrelevant distractions in this exciting and epoch-making transformation driven by private media, communications and property interests. Consequently, little thought is being put towards how urban policies, plans and strategies can *engage with* new technologies as policy agents to try and help shape *desired* urban futures in a purposive manner.

With utopianism and crude technological determinism often dominating popular (and, in many cases, academic) debates, it is not surprising that the potential roles of urban policy makers and planners in 'socially shaping' new technologies in cities at the local level are usually overlooked (see, for example, Negroponte, 1995; Martin, 1995). This neglect, however, is problematic. It means that a fast-growing wave of urban experimentation with telecommunications, which is currently emerging across advanced industrial cities, is largely ignored within urban planning and policy debates. This is a problem, because such innovation promises to have major practical and theoretical implications for how we might consider the future of cities, urban policy and planning. It may also offer lessons on the broader question of how we might best understand the relations between cities and new communications technologies, and how we might address the crucial question of thinking about the ‘local’ and the ‘urban’, in an increasingly tele-mediated and globalised era.

This paper attempts to address this problem by inserting urban planning into debates about new technologies and the future of cities. It has three parts. In the first, we set the context for local, planned intervention on IT and telecommunications by reviewing the evidence on the complex interdependence between cities, transportation and face-to-face interactions, and electronically-mediated interactions. We do this by looking in turn at transport-telecommunication relations, the broad links between urban economies and the so-called 'information economy', the relations between urban and so-called 'cyber' cultures based on IT systems, and, finally, the ways in which urban communities and 'virtual,' IT-based communities are interdependently linked. In the second part of the paper we go on to review a broad, international range of emerging policy initiatives which aim to help shape the articulation between the built form and socio-economic development of cities, and the electronic interactions within such cities, and between them and the 'outside' world. Finally, in section three, we assess the significance of these policies and suggest ways for creatively integrating telecommunications into urban policy and planning practices and strategies.

The Metropolitan Dominance of the 'Digital Age'

To understand why cities will be central to the 'digital age', we need to explore the *complex inter-relationships* between electronic and urban interactions in the economy, culture, and society. We need, in short, to understand how our urban life stands in a state of subtle, two-way, *articulation* with electronic interactions (see Robins, 1995). We also need to understand why the level of the city and urban region might be a crucial one for exploring new policy innovations which make the most of the potential of new technologies, in a way which has meaning in relating to the urban worlds in which the vast majority of us live, work and act out our lives.

To further explore the co-evolution of cities and telecommunications, which is essential as a basis for successful urban policy and planning innovation in these areas, we need to look in more detail at its four key aspects: transport-telecommunications interactions; the links between urban economies and the 'information economy'; interactions between urban cultures and cybercultures; and the subtle interactions between place-based and 'virtual' communities in cities.

Transport / Telecommunication

Conventional approaches to transport - telecommunications relations stress the environmentally beneficial role of telecommunications technology. It is often assumed that telecommunications can unproblematically substitute for physical transport flows and movement, reducing the need for travel, and so lowering levels of pollution and urban congestion. For example, BT's Environmental Manager argues that: "telecommunications technology is likely to play an increasingly important role in offering a more environmentally sound alternative to travel ... Apart from a saving in energy the switch to telecommunication services would have other environmental benefits such as reduced noise levels, fewer new roads and lower levels of urban pollution". Jonathon Porrit, one of the UK leading environmentalists, "hails this fusion of communications and computing technologies as one of the 'tools for sustainability".

Early research on the potential for trade-offs between telecommunications and transportation networks was commissioned by the US government in the mid 1970s in response to the energy crisis. This work simply compared the energy costs associated with communications through the telephone, and physical forms of communication such as personal travel by car, train and aeroplane (Nilles et al, 1976). The energy savings associated with communications by telecommunications created much excitement about the potential for trade-offs between telecommunications and transportation.

Initial interest in the role of telecom - transport trade-offs focused on the potential for the displacement of work-related commutes. In the mid 1970s, Jack Nilles invented the term 'telecommuting' to describe home or neighbourhood-based working using computers and telecommunications technology (see Nilles et al, 1976). Tele-based communication was seen as a solution to the problem of congested urban environments and long commutes to centralised offices. It was assumed that telecommunications would simply substitute electronic flows for the transportation of people and freight along more polluting road, rail and air networks. There were a number of attempts to demonstrate the potential substitution effects of teleworking on travel patterns and estimate energy savings. In the US context, these demonstrate that telecommuting has the potential to save between 1 -3% national energy consumption, figures which were not as great as the early proponents of teleworking might have expected (Nilles, 1988). These reports also have highlighted some of the "rebound" implications of teleworking including extra energy consumed in home, movement out to higher-amenity areas, increasing total commute distances and the rapidly filling of road space through high latent demand (see Marvin, 1997).

The UK Department of Environment, Transport and the Regions' traffic projections show how demands for mobility and movement, both within and between cities, are unlikely to reduce, even whilst telecommunications use continues to burgeon. Overall, transport and telecommunications *actually feed off and fuel, more than simply substitute, each other.* While telecoms undoubtedly have some potential to substitute for journeys and more routine interactions there is considerable evidence that the relationship between transport and telecoms is more complex (Graham and Marvin, 1996). Rather than simply substituting, telecoms have highly *complementary* relationships with physical travel. This, we would argue, can actually result in three forms of traffic *growth.*

First, telecommunication play a central role in *improving the efficiency and effectiveness of transport networks,* so reducing the cost of travel. Complex computer ticketing, transaction systems and air-traffic control systems help reduce the costs of air travel, making it more attractive as our perceptual understanding of the world increases. Electronic information exchange also plays a major role in the organisation and management of transport networks. Such innovations help to extend the reach, reliability and usefulness of transport flows through air travel, auto-route guidance, fax, mobiles, e-mail, real-time information, and electronic data interchange-based 'Just-in-Time' logistics systems. A single flight of a 747, for example, has been estimated to generate 50,000 electronic exchanges in booking, maintenance, refuelling, airport management etc.

Second, access to cheaper telecommunications increases an individual's or organisation's 'perceptual space', creating more opportunities for physical travel. The more we get to know about the world, whether from a leisure , recreation, or business opportunities point of view, the more we demand new forms of physical travel to support flows of goods or to directly experience the quality of interaction that can only be achieved through physical contact -- usually in cities. The people with the greatest demands to use phones and mobile computers are business travellers. People who telework and move far from cities may actually travel further overall because they travel further for other trips like shopping or because there fewer commutes are over longer distances.

Finally, congested roads create new demands for telecommunications. Mobile phones may actually help to sustain larger traffic jams because they allow 'dead' time to be converted to 'live' working time. It is no accident that some car manufactures now sell cars with car phones, faxes, or mobile computers - the ideal way of staying in touch with work and home, once a driver is stuck in a gridlock or slow moving traffic. In this way telecommunications helps overcome our resistance to travelling by cars on roads which are congested, as real time information overcomes many of the uncertainties and difficulties of travel. In short, what appears to be happening is a major expansion in all forms of communications. Although some substitution may undoubtedly be taking place, overall growth of electronic and physical mobility simply overwhelms the contribution of substitution.

Urban Economies / Information Economy

Current advances in telecommunications are a phenomenon which is overwhelmingly driven by the economic dynamism of cities, particularly larger, internationally-oriented, metropolitan regions. City-regions have important assets in an internationalising economy, based more and more on flows of information, services and 'symbolic' products like media, advertising, cultural services, and electronic entertainment (as well as movement of people, goods, and commodities). They support face-to-face interactions, especially for higher-level decision-making functions, in a world of fast flows and great volatility. We should not forget the sheer infrastructural advantages of cities. Cities also have the high-quality physical, service and telecommunications infrastructures to extend access efficiently to distant places and markets. Whilst remote, rural areas might still have old-fashioned and poor quality analogue telecoms infrastructure of the old monopoly (BT), our main city centres now have three or more separate, high capacity digital systems competing in price and quality (with many others selling specialised services). London's City and West End districts has at least four superimposed optic fibre grids (those of BT, Cable and Wireless, MFS and COLT) and countless other service providers who deliver over these networks. 80% of investment in telecoms in France goes to the Paris city-region.

But cities still offer unrivalled place-based, as well as electronic, contact potential. Today's uncertain and globalising economies make trust, constant innovation, and 'reciprocity' more and more important, which can only be fully-forged through on-going face-to-face contact. Stressing "the extraordinarily social nature of modern economies," the geographers Nigel Thrift and Kris Olds (1996, 316) write that "it is clear that face-to-face interaction has not died out. Indeed, in some sense it has become more important as reflexivity (including and enhanced ability to see oneself as others see us) has become built into economic conduct". Tony Fitzpatrick, (1997, 9) the Director of Ove Arup, argues that "cities reflect the economic realities of the 21st century. Remote working from self-sufficient farm steads via the Internet cannot replace the powerhouses of personal interaction which drives teamwork and creativity. These are the cornerstones of how professional people add value to their work. Besides, you cannot look into someone's eyes and see that they are trustworthy over the Internet". On the consumption side, too, the whole range of consumer services, now so important in urban economies -- tourism, shopping, visiting museums and leisure attractions, eating and drinking, sport, theatre, cinema etc. -- are all growing and seem likely to resist any simple, substantial substitution by 'on-line' equivalents.

Major urban places support dense webs of face-to-face links, transactional opportunities, agglomeration economies, access to wide pools of specialised labour, services, property and infrastructure, and 'soft' cultural and social advantages. Here the emphasis is increasingly on the *qualitative* aspects of urban economies and urban places, and the increasing dominance of urban economies by symbolic and representational flows and outputs, which may or may not be linked closely to commodity flows and outputs (what Lash and Urry (1994) term 'economies of signs and space'). Thus, in today's urban economies, the spiralling use of telecommunications becomes combined with the spiralling use of transport and a continuing, perhaps even growing, reliance on face-to-face contacts and meetings, largely in cities: what the father of the 'Megalopolis' concept, the geographer Jean Gottmann, used to call the "spiralling mass of bits of information". What many people interpret to be a post-urban shift might actually be a transition from traditional, core-dominated, monocentric cities towards complex, extended and polycentric city-regions made up of a multitude of superimposed clusters, grids and internal and external connections.

This is an important argument, which leads us to stress four points. First, growing flows of electronic information may require *more* face-to-face contact to make sense of it all (as the Bristol geographer, Nigel Thrift, has shown with his work on City of London electronic traders). Second, much of the electronic exchange on networks like the Internet actually represent cities as places to visit, consume, travel to, or live within, as the many tourist and municipal and urban services sites on the Web demonstrate. In the US, the fastest-growing web sites are those that try and integrate the Internet at the level of the metropolitan region simply because this has most salience to Net users. Private operators of integrating city-level web sites in the USA have grown massively recently, attracted by the increasing maturity of the web as a diffused medium and by their calculations that "80% of purchases are made within a 20 mile radius of the home" (McElvogue, 1997, 2).

Third, it is easy, given the hyperbole about globalisation, to radically underestimate the degree to which tele-mediated flows operate to sustain very local relations. The vast bulk of electronic exchanges, for example, are very local indeed: 60% of all phone calls and e-mails move within single buildings (Graham and Marvin, 1996). And, finally, even when stories are apparently about decentralisation and substitution, essentially urban dynamics are usually at play. Most teleworking, for instance, is done for part of the week in the zones within and around the large cities that allow people to go to the office on one or two days a week for face-to-face meetings. When IT does support so-called 'dis-intermediation', directly linking consumers and producers across distance -- as is the case with many call-centres delivering routine services on-line -- their destinations tend not to be rural spaces but smaller, provincial, cities (or, in sectors like on-line computer programming, far-off cities like Bangalore in India).

Urban Cultures / Cybercultures

Thirdly, the centres of many of our larger cities are experiencing renewed growth as interlinked centres of growing cultural industries (arts, theatre, dance, music, publishing, fashion, media, graphic design, photography, architecture, leisure, sport etc.). The importance of city centres has recently been re-emphasised, based on the widening assertion that such 'cultural' industries' may, with appropriate policies, interlace positively within a framework of public space to support the emergence of "creative cities" (Landry and Bianchini, 1995). The central idea driving such policies is that cities can only thrive when strategies recognise that "the defining characteristics of cities are high density, mixed use, stimulus, transactions and above all diversity" (Montgomery, 1995). But how do urban cultures interrelate with much-vaunted 'cybercultures,' at a time when *electronic* cultural interactions are exploding, fuelled by digital technologies, and the blurring of boundaries between computing, media, and telecommunications industries and technologies?

It is clear, at the level of large organisations, that large cities *already tend to dominate* the sectors that are blurring together with the emergence of multi-media: TV, publishing, art and design, film, and media and telecommunications. Such firms continue to rely on large cities for all the reasons stated above. But the prospects for creative synergies between urban and cybercultures seem extremely strong at the much more dynamic level of small and micro-firms. In an apparently paradoxical twist, the continuous innovation in the Internet, digital media, and multimedia content industries, so often hailed as supporting the 'death of distance' by media and industry pundits, is being fuelled by intensely local networks based on face-to-face interaction in selected urban districts.

In fact, as the value-added in IT industries shifts from the zones dominated by hardware production to places that can sustain innovation in software and content, so the focus of the industries may actually be shifting from Silicon--Valley--like Research and Development campuses, to central, old-city, locations. In the cultural industries, the creative small firms that dominate Internet software, digital design, and World Wide Web services, far from scattering towards rural idylls, seem, in fact, to be concentrating into (a small number of) gentrifying metropolitan 'information districts' like SoHo and Tribeca in New York, Shoreditch in London, and Temple Bar in Dublin. As well as having good (broadband) telecom connections and tailored, 'Internet-ready' office spaces, such districts are thriving through processes which, arguably, are analogous to those which spawned the first industrial districts in 19th century cities. In a detailed study of SoHo and Tribeca in Manhatten, for example, Dan Hill (1997) found that the raw material for such industries is the sort of informal networks, high levels of creativity and skill, tacit knowledge, and intense and continuous innovation processes that become possible in an intensely-localised culture, based on on-going, face-to-face contacts supported by rich, dense and interdependent combinations of meeting places and public spaces. Clustering in certain 'information districts' may thus support the informal and on-going innovation networks and serendipitous contacts that seem central to the success of small and micro digital arts and creative firms (Hill, 1997)

Most importantly here, the Internet, with its 'spiralling mass' of information, communication, transactions and specialised media flows, is now weaving into support every aspect of urban functioning in contemporary city-regions. Such trends are most advanced in the USA, which demonstrates the strong metropolitan bias of both the production and consumption sides of the Internet. On the *production* side, for example, down town New York's 'Silicon Alley' has emerged, along with districts in downtown San Francisco and other large cities, as a remarkable concentration of micro and small firms, based on digital art and design, Web production, and digital and multimedia services. These draw on the City's unparalleled arts, cultural industries, and literary traditions. One of the main motors of the recent economic renaissance in Manhatten, Silicon Alley encompasses over 700 new media firms who rely on intense, informal, local contacts to sustain continuing innovation and interaction (Hill, 1997). Interestingly, urban planning and policy is beginning to find ways of supporting this new information district, as we shall see in the next section. It must be stressed however, that currently only a relatively small number of urban districts are being redeveloped in this way. Moreover, these 'organic' spaces are very difficult to develop 'artificially' - that is, in the absence of a existing appropriate, high skill levels, contact networks, supportive services, and local entrepreneurial culture.

On the *consumption* side, it seems that the metropolitan dominance of the Internet might actually grow rather than decline, as it becomes massly diffused, accepted and gradually weaves into all aspects of urban life. This is certainly the recent experience of the world's most mature Internet market, the USA (Moss and Townsend, 1997). The top 15 metropolitan core regions in the US in Internet domains accounted for just 4.3 % of national population in 1996. But they contained 12.6% of the US total in April 1994. By 1996 this had risen to almost 20%, as the Internet became a massly-diffused and corporately-rich system. As Moss and Townsend (1997) suggest, "the highly disproportionate share of Internet growth in these cities demonstrates that Internet growth is not weakening the role of information-intensive cities. In fact, the activities of information-producing cities have been driving the growth of the Internet in the last three years ". Manhattan now has twice the 'domain density' (i.e. concentration of Internet hosts) of the next most 'Internet-rich' US city - San Francisco - and six times the US average (Moss and Townshend, 1997).

In such a context, there are many opportunities for weaving "access points" - Internet and service kiosks - into the fine-grained fabric of cities, to animate, enliven and inform what goes on in the public and private realms of cities and metropolitan regions. Whilst there are many problems here to do with the high costs on technologies, highly uneven social access and skill levels, and dangers of oppressive surveillance, a growing range of initiatives at the urban level are currently experimenting with new media solutions to support the improved delivery of public services, support community networking, and enhance local economic, social and cultural development.

But such public initiatives are far outweighed by the sheer *economic* logic of the Internet. Already, private Internet providers are themselves starting to develop integrating web sites at the urban level. These aim to support coherent and legible relationship between the many services on the web that fall within a particular city and the population of that city. The need for 'urban Internet guides' is especially powerful in the larger 'global' cities that dominate Internet innovation and use. New York, for example, how has over 10 dedicated 'virtual city' web sites that aim to draw together various portfolios of local Internet-based services (Graham and Aurigi, 1997).

Urban Communities / Virtual Communities

Finally, all aspects of the social use of telecommunications remain highly dominated by, and bound up with, the lives and social worlds of urban populations and communities. It is in our metropolitan regions that the most rapid diffusion of mobile phones, cable systems, and the Internet, have developed, and where the rich communicational and transactional fabric of cities is increasingly supported by complex tapestries of telecommunications networks. One only has to witness the recent explosion of the use of mobile phones on our city streets to understand this. Thus, for example, the many virtual communities on the Internet are made up of both globally-stretched Use-net groups and Multi-User Dungeons (MUDs) etc. and a growing range of community networks at the local level in towns and cities, aimed at feeding back positively onto the social dynamics of individuals cities (Graham and Aurigi, 1997).

Some writers have even suggested that local urban community networks like the Cleveland Freenet, Santa Monica Public Electronic Network and Seattle Community Network, as well as the more recent 'virtual ' or 'digital city' movement, represent hope for truly interactive, democratic, media systems which might help revive, enliven and inform the public realm of their host cities (see Schuler, 1996). The hope here is that local IT systems that support interactive community debates, will help to bring together the diverse social, cultural and geographical fragments of extending city regions, adding important coherence and legibility to a city's 'electronic realm' in the process. In the long run, computer networks 'grounded' in particular local communities, might be more sustainable, effective and meaningful than those which rely purely on IT-based exchange across global distance. Wakeford (1996) argues that 'grounded' community IT networks can often support higher degrees of trust (with "persistent" rather than "transient" identities). They can also relate more effectively to real problem-solving, can inter-relate with training centres and face-to-face contact, supporting reciprocal, frequent and supportive interactions which relate strongly with the wider public realm. Finally, these advantages can often allow them to draw in wider cross-sections of people than global Internet newsgroups.

But such assertions and rhetoric need to be tempered by the realisation that profoundly deep social divisions in access to all communications technologies are woven deeply into the fabric of our cities (see, for example, Demos, 1997). Whilst elite groups are 'super connected' to phone and IT networks at home, school, in their cars and at work, even the humble telephone is an expensive luxury in many more marginalised urban neighbourhoods. One neighbourhood in inner Newcastle, for example, had only 27% of its households connected to the telephone in the late 1980s (Graham and Marvin, 1996; Chapter 5). Home access to the Internet, with its pre-requisites of skills, electricity, space, hardware, software, telephone, modem, Internet account, and cash for on-line and phone charges, is unlikely to be a priority for the large proportion of socio-economic groups facing poverty, debt, and problems paying for essential bills. This places a premium on supporting access to community IT networks and Internet-based systems into the public spaces of cities.

One of the key policy issues for the tele-mediated city, therefore, is how to address deep social segmentations based on access to, and exclusion from, the new communications media, and the growing ranges of information, resources, and transactional and working opportunities offered over them (see Schön et al, 1998). As more and more IT-based systems become the norm -- for example with growing reliance on home and mobile telephones, home IT systems, and electronic cash -- so it becomes more and more disadvantageous to experience 'network poverty' beyond the reach of such systems. But this is not to assert that systems like the Internet can act as some 'silver bullet' to complex problems of social exclusion, in the manner of the US Speaker, Newt Gingrich's absurd, utopian, and patronising "laptops to the ghettos" rallying cry in 1995. Rather, it is to stress that new community-based electronic networks are required which, as a matter of course, work to enrol the broadest possible range of users and voices. Such experiments must be seen as attempts to explore the fullest potential of the new media, as realms for social communications at the most meaningful geographical level to most people - that of the metropolitan region.

Emerging Telecommunications-Oriented Urban Planning Strategies

Clearly, it is no longer adequate to consider policies for cities and those for telecommunications and new media entirely separately. The above imperatives suggest that only through addressing the complex interactions between cities and telecommunications will the potential of the technology be realised. This realisation is currently leading to a wide range of policy experiments which aim to positively shape how the new media relate to specific cities or parts of cities. Early examples can be drawn from a 'bricollage' of evidence, drawn from cities in the US, Canada, Malaysia, Europe, the UK and elsewhere around the world. These examples have not yet coalesced around a coherent new paradigm of urban policy. Many can be criticised as technologically determinist, environmentally problematic, or socially exclusionary.

But we would argue that, together, they point to a new style of planning and urban policy. Such urban strategies try to shape face-to-face interactions in place (and the transport flows that sustain these) in parallel with electronically mediated ones across distance. Currently, we can identify three emerging styles of such 'urban telecommunications planning' : integrated transport and telecommunications strategies, city-level new media strategies, and so-called 'information districts' and 'urban televillages'.

Integrated Transport and Telecommunications Strategies

The first set of initiatives attempt to shape and manage the relations between physical movement and mobility through the application of new media, combined within particular forms of urban physical development. Each embodies a particular conception of the relations between different forms of communication and their role in the development of the city.

Urban and Regional Teleworking Initiatives

First, there are Urban and Regional Teleworking Initiatives. There are a growing set of initiatives, especially in the USA, that are attempting to grapple with the problems of developing a metropolitan-region approach to teleworking to make a positive contribution to environmental improvement - particular reducing vehicle emissions. For example, Telecommunications for Clean Air is a two year programme funded by the Californian South Coast Air Quality Management District to use telecommunications to meet rigorous air quality standards. The main aims of the programme are to identify cost-effective solutions to air quality and congestion, to contribute to economic growth, and to develop a regional approach to problem solving.

The Telework Facilities Exchange is designed to expand telecommuting participation in local government by providing low-cost, flexibly-located facilities and marketing these practices to other organisation joining the exchange (Telecommunications for Clean Air, 1994). A public sector employee would normally commute 35 miles each way to their office. Instead, they commuted to a vacant office a few miles from their home to use a workstation connected to their office. Those workers participating in the Programme reduced their normal vehicle miles travelled by 88%. As a result, if 30% of the region's 484,000 local government employees and each worked one day a week at the exchange, nearly 500,000,000 vehicle miles travelled could be saved each year (Telecommunications for Clean Air, 1994).

These findings have to be treated with caution because studies indicate that teleworking can generate "rebound effects". Although commute trips and time can be saved, additional recreational and shopping trips may be generated. Telecommuters may also decide to live further from the city, so increasing the length of remaining commutes. And the space freed on highways is quickly filled with new commuters. (see Mokhtarian, 1990; Department of Energy, 1994). But these tensions could be managed at a metropolitan level, as teleworking is coordinated within the context of wider transport and landuse strategies. The US initiatives are particularly interesting because of the high degree of organisational innovation in the delivery mechanisms for teleworking, and the much stronger links to more mainstream transport, air quality and landuse planning policies than tends to exist in Europe.

Assessments of the environmental potential of teleworking in the UK have largely remained at the level of national aggregate assessments (British Telecom, 1992; CEED, 1992). Overall, the conclusions are that the beneficial impact of telecommunications may be limited. Even assuming that teleworking continued to grow at a fast rate, the Royal Commission on Environmental Pollution concluded that it would only have a small role to play in the reduction of emissions. But, although the national environmental benefits may be small, there could be potential in large conurbation's in the UK to develop initiatives within a wider environmental policy framework. A recent study by a Cambridge-based consultant, for example, argued that 1.25 million miles year could be saved by 2000 office-based staff at Cambridge council telecommuting (Environment, 1997). Further development work is necessary, perhaps drawing critically on the US experience. For instance, BT provides limited guidance for local authorities attempting to integrate telecommunications into Agenda 21 Strategies. And the Association of County Councils has started to consider how IT and telecommunications could be linked with the green agenda (BT, 1997; Association of County Councils, 1996). The key issue is the need to carefully integrate teleworking within the context of wider transportation and landuse plans.

Communication Corridors

Second, there are new forms of communication corridor strategies which attempt to shape how telecommunications, transport and land use interplay within broader urban commuting corridors. Those based on existing rail/transit networks attempt to manage travel demand both on the road and rail network through the provision of teleworking centres and incentives to travel off-peak. The Metro Net initiative in Los Angeles, for example, involves retrofitting a high capacity fibre optic network along side the 300 mile regional Metro Rail and Link network (Siembab, 1992). The proposals are designed to achieve three objectives: to generate revenue through leasing capacity, to develop services to enhance ridership of the system, and to improve regional mobility through developing station-based employment and service centres. The mobility strategy has been designed to fit in with the objectives of a series of wider regional communications and landuse strategies. More specially, it is hoped that the strategy will contribute towards the implementation of the Air Quality Management Plan, the Development Plan, the Regional Congestion Management Program, and the promotion of regional economic growth. The proposals would develop telecommunications facilities at or near stations for conferences, education, and job-training, to make the Metro system a destination itself.

Further development would link both the metro and telecoms networks to targeted parcels of adjacent land, in order to attract new employers whilst maximising public transport usage. These ideas mirror proposals in the UK to set up telecommuting offices along the Folkestone to Waterloo rail line, in order to reduce the number of rush hour commuters and enable a channel tunnel link to run on existing tracks (Roarke Associates, 1994). A more recent proposal suggested building a ring of teleworking centres around the M25 to deal with predicted car traffic growth. The architects, Roarke Associates, argued that the teleworking centres would cost £450 million whereas road widening would cost £1450 billion - a saving of £1 billion (*Times*, 1997). Both these initiatives are at the proposals stage and have not been implemented. However, they illustrate some interesting ideas about how communications planning could integrate telecoms and transport, and start to make wider links to landuse strategies. Yet, at an institutional level, there is still considerable uncertainty as to which organisations could take the lead with such complex initiatives involving so many different dimensions of planning.

Road Transport Informatics (RTI)

The third type of initiative focuses on the development of Road Transport Informatics (RTI). City-wide initiatives are rapidly emerging here, concerned with the use of RTI systems to more effectively manage transport networks. There are major initiatives in Europe, the US, and Japan and the National Economic Development Council estimate that the global IT and traffic management market will be worth £29 billion in 2010. However, objectives of RTI are often poorly defined and are not often closely linked with landuse and work patterns. These initiatives are more often characterised by a form of 'technical fix,' dominated by strong producer-led interests. RTI strategies have assumed importance in the context of EU funding programmes where the technologies are seen as making a major contribution to sustainable development (CEC, 1995).

There have been a large number of feasibility schemes evaluating the potential of various forms of RTI and electronic tolling in the UK. But, again, there has been a failure to link such debates within the context of wider urban management, regeneration, and landuse strategies. These issues are being considered more widely in North America. The Highway 407, which is currently being built in Toronto, is billing itself as "Tolls but No Jams" (*Toronto Star* July 29, 1996 pages A6-A7). Located in one of the most congested highway corridors in North America, a $1 billion dollar 36 km highway will eventually connect the airport to downtown Toronto. The scheme is being developed by Canadian Highways International Corporation -- a private consortium of 4 companies who will be funded by the receipts from electronic tolls. In return for the higher charges, users will benefit from higher road speeds than the current limit, and no traffic jams. If demand increases, the highway can be expanded to 10 lanes and tolls can be raised to reduce peak travel volumes. The scheme is being marketed at those firms operating just-in-time production methods who require high degrees of certainty in travel times for the movement of goods and services. There is now major development interest in highway land involving commercial, retail, leisure and recreation and housing adjacent to the new road. This initiative is an interesting example of the combined planning of electronic, transport and landuse infrastructure, designed to develop a congestion-free, higher-speed and lower travel-time corridor through the congested region. But this new development trajectory is extremely socially exclusionary. It is very much designed to meet the needs of large international corporations and elite users prepared to pay the premium for increased certainty.

Other RTI initiatives focus on the development of local and regional initiatives in driver information and control systems. These initiatives are based on proposals to carry more traffic by making better use of the existing road network, through pre-trip planning, route guidance, traffic management and control, and network management applications. It is hoped that the provision of information to drivers on road conditions can increase the efficiency of the network to minimise delays, unreliability and environmental damage. For instance, it is estimated that driver information could increase the capacity of the road network by 1% provide a 10% saving in journey times and a 6% reduction in mileage. The Scottish consultative document argues that the region is well-placed to use these technologies because 80% of the population live within a relatively self-contained belt across the country (The Scottish Office, 1993).

There are a number of problems with the emerging set of urban communications strategies in the UK. They tend to remain largely disconnected from mainstream urban strategies and have poorly developed links with landuse planning. They also tend to narrowly represent a limited and technologically-determinist view of urban futures based around notions of a producer-led 'technical fix' to the problems of urban mobility. Nevertheless, taken as a package, they do begin to illustrate some of the ways in which integrated urban planning for physical and electronic mobility.

City-Level New Media and IT Strategies

The second broad emerging policy area is City-wide new media strategies. IT strategies for community networking, local economic development, and public service delivery have been underway in many UK, European, and American cities for a decade or more. Following American experience, community networks like Free-nets, the Manchester, Kirklees, and Nottingham Hosts, and the Newcastle NewNet system, based on the Internet, have emerged which try and use computer communications to support grassroots, local economic, and voluntary activities (see Graham and Marvin, 1996). Many local authorities are also experimenting with videotex systems, electronic kiosks and smart cards systems to deliver information on public services, and aim to improve the services themselves. Nationally in the UK, the new government report, *The New Library: The People's Network*, proposes to wire up both schools and libraries as places where IT networks can be made widely available for local communities.

Local services have developed patchily on the new urban cable networks developing across the UK. And virtually all major UK cities now have a presence on the Internet, where so-called 'virtual cities' range from simple tourist promotion and local databases, to sophisticated spaces which attempt to add coherence to all local activities on the Internet, to widen local access and skills, open up interactive services for local debates, and to develop information and communications services which feed back positively onto the development of the home city (Graham and Aurigi, 1997). Interestingly, the most innovative virtual cities use the analogies of city 'spaces', 'squares' and 'districts', so that the many services they offer relate directly to their counterparts in physical urban space. The most sophisticated on these in the UK currently is Virtual Bristol, supported by a partnership of the City Council, Universities, and Hewlett Packard, launched in April 1997. Not to be left out, BT is exploring the concept of "urban intranets" - Internet services that are only accessible to specified local communities.

This disparate range of local new media initiatives have two problems, however. First, they have tended to be fragmented local 'IT islands', largely ignoring each other. And second, they have usually been developed with little or no respect for how they relate to the physical urban realm or to the broader development dynamics and geographies of their subject cities. Thus, the challenge for UK cities is to shape coherent partnership-based strategies aimed at harnessing all types of new media applications - Internet, cable, kiosks, telephone, infrastructure - to their economic, social and cultural development needs. Such issues need to begin with social, geographical and institutional issues and policy needs and move onto how new technologies might meet these needs - rather than the other way round. Institutional solutions need to be found that harness the entrepreneurial energies of the new media industries, and their growing interest in market-based local initiatives (like the booming commercial metropolitan Internet sites in the US), whilst linking creatively and positively to the fragmented sets of agencies involved, in the broadest sense, in the governance of UK cities (local authorities, development agencies, health, education and information institutions, firms, schools, the community and voluntary sector etc). Clearly, urban media 'Master plans' will be impossible : what is needed are strategic frameworks so that the innumerable local media investments and initiatives emerge to be more than the sum of their parts.

Finally, and most importantly, from the point of view of this paper, there needs to be a much more thorough-going attempt to link urban media strategies to the development of cities themselves, so ensuring that, wherever possible, synergies can be developed between media and place-based exchanges. Progress is being made here, however, at both the urban and regional levels. At the city level, strategic planners in cities like Amsterdam and Lille have already attempted to integrate new media into the future urban visions. In dozens of cities across the world, 'teleports' and 'telezones' have been designated in particular urban districts blending advanced office and business space and sophisticated telecommunications facilities. In the US, ambitious 'Smart City' new media strategies are tentatively starting to consider land-use planning and urban policy issues - as our discussions of strategic urban corridors and information districts demonstrate. Already, in some such (highly affluent) communities like Palo Alto and Blacksburg Virginia, very high levels of Internet and e-mail access and use have begun to transform the communicational fabric of urban areas as these new media become woven into the fabric of urban life (see Graham and Marvin, 1996) Predictably, in a private-sector-led planning process, the consultancies engaged in 'Smart Community" planning argue that "cities unprepared for these [new media-based} changes risk being consigned to geopolitical obsolescence before they even know what hit them" (International Center for Communications, 1997).

In the UK, the packaging of IT infrastructure with individual land-use developments - business parks, 'telecottages', 'wired villages' etc - is increasingly common. But efforts are also starting to link broader urban media strategies with urban-wide development strategies. After a period when grant, training and technological support was 'pepper-potted' through the city, Manchester is increasingly gearing its broad telecoms and IT initiatives to specific urban redevelopment and re-use projects, and to strategic discussions about combating social exclusion in the City. A widening range of new physical, IT-oriented projects have emerged linked into the network services on offer : the Electronic Village Halls (linked to community centres and initiatives through the City), existing managed workspaces in New Mount Street, and proposed ones in the 'Northern Media Quarter', and a centre for Multimedia development and applications in Hulme (Carter, 1996). A similarly broadly-based IT and new media strategy, known as the GEMESIS project, is underway in adjacent Salford, backed by a broad partnership between the cable company, IT firms, local universities and training providers. In partnership with its university sector, Manchester/Salford, like German cities such as Berlin and Bochum, is also building a new broadband Metropolitan Area Network ('MAN') infrastructure ring in the City that will spur efforts to regenerate the inner city through research and development and scientific innovation.

Europe-wide, cities at the vanguard of new media strategies are co-operating through the 'Telecities' network to exchange experience, develop lessons for best practice, support pilot initiatives, and lobby the EU for further support. Telecities also links into Europe's efforts to support the emergence of what it terms a 'Regional information Society' through its broadly based Inter-Regional information Society Initiative (IRISI). This supports integrated packages of ICT based pilot projects in designated regions in sectors as diverse as education, health, social services, transport and logistics, media and public services. But, again, there exist few links between the way the EU is considering ICT applications, services, and infrastructure, and its broader considerations about its spatial development in the future (CEC, 1997).

'Information Districts' and 'Urban Televillages'

The final area where new media policy is becoming directly linked with policies for particular urban spaces is the emergence of 'information districts' and urban 'televillages". Building on the debates about 'urban villages' in Europe, and the 'new urbanism' movement in the US, interest is growing rapidly in how media infrastructure and services can be designed and managed, geared to sustaining and feeding back on particular urban districts. In California, the concept of the 'TeleVillage' -- an integrated urban place supported by a whole suite of ICT infrastructures and services -- is gaining support. The Blue Line TeleVillage, a two square mile area on one of the new public transit corridors in LA, is based on a holistic strategy to manage land use, transport trips and electronic communications so that synergies emerge between the three, creating a 'liveable' community with reduced automobile use, higher community-based activities, and higher urban densities than in the usual LA suburbs (Siembab, 1995).

Physical places for supporting IT training and services - community centres, computer centres, telework centres, IT links in schools, hospitals, transport facilities and libraries, and electronic kiosks in public and semi-public spaces - are integral to the plan which is backed up by a broad, public-private-community partnership, and an extensive array of on-line public services. In partnership with the public transport operators in the LA region, a new fibre network is being developed to link together whole constellations of TeleVillages across the region.

Different packages of IT and telecoms infrastructure and services are being offered for different land uses; 'distributed' organisations are being encouraged ; and attempts are being made to include more marginalised social groups. The philosophy is that IT-based retrofitting in existing US urban areas will mean that many urban problems might be addressed "with very little new physical construction and no dramatic changes in density" (Siembab, 1995).

The other emerging example of combining new media and urban regeneration at district level is the concept of the 'information district'. Here, the emphasis is on creating urban 'milieux' that sustain economic growth in new cultural and 'symbolic' industries, where informal face-to-face contact is essential, whilst also providing high capacity on-line linkages to the wider world. Such strategies are inspired by the emergence of 'information districts' described above. Most often, information district strategies emerge organically, as in the cases considered above - New York's Silicon Alley, Dublin's Temple Bar, and Manchester's Northern Quarter - where clusters of such industries emerge spontaneously in inner urban districts. Then, the challenge is to intervene to further support the growth of small and micro firms in the relevant sectors, whilst also ensuring appropriate property is available and that broader efforts are made to improve the broader urban realm and the contribution of the industries to the economic and social revitalisation of the city as a whole (Hill, 1997). Thus, both New York and LA have offered grant schemes and tax exemptions to small and micro firms in the new media sectors. Backed by the powerful New York New Media Association (NYNMA), specialised multimedia centres, offering managed workspaces and high-level telecoms bandwidth, have also started to emerge in Silicon Alley, as have dedicated venture capital funds and orchestrated events and programmes designed to encourage local face-to-face networking. Elsewhere in the USA, the city of Spokane, in Washington State, has wired up much of its downtown to attract multimedia firms.

In Europe, strategies at the neighbourhood and district level have begun to look to coherent interventions in the urban realm, new media, and, at the institutional level, to try and either sustain, develop or encourage, local clusters of multimedia firms Manchester has explicitly adopted the Silicon Alley model to support its Northern Media Quarter, on the edge of the city centre. Sheffield's well known Cultural Industries Quarter (CIQ) strategy, aimed at clustering the broadest possible range of media, design, music , film and cultural-related industries firms in one part of the City centre, is now backed-up by widening range of on-line services financed by a public private partnership called NEO (Hill, 1997). Dublin's Temple Bar district is backing up its physical regeneration efforts, weaving a parallel infrastructure for electronic, multimedia exchange.

In London's Soho media core, mean while, a specialised telecommunications network was recently constructed by a consortium of film companies called 'Sohonet'. This system links the tight concentration of film and media headquarters in the district directly to Hollywood film studios, allowing on-line film transmission and editing over intercontinental scales, via highly-capable, digital, broadband connections. The network is seen as a critical boost to the broader global ambitions of the UK film and cultural industries.

Conclusions: Integrating Telecommunications into Urban Planning

In this paper we have attempted to demonstrate that new information technologies actually resonate with, and are bound up within, the active construction of urban places, rather than making them somehow redundant. Urban places and electronic spaces are increasingly being produced together. The power to function economically and link socially increasingly relies of constructed, material places which are intimately woven into complex media infrastructures which link them to other places and spaces. "Today's institutions" argues William Mitchell (1995; 126) "are supported not only by buildings but by telecommunications and computer software". Thus the articulation between widely-stretched media and telecommunications systems, and produced material places, becomes the norm. It is, indeed, a defining feature of contemporary urbanism. "Constructed spaces", continues Mitchell, "will increasingly be seen as electronically-serviced sites where bits meet the body -- where digital information is translated into visual, auditory, tactile or otherwise sensorily perceptible form, and vice versa. Displays and sensors for presenting and capturing information will be as essential as doors" (Mitchell, 1994). As cities extend into polycentric metropolitan regions, the spaces of the city are being constructed within broader, and more complex, urban fields, networked together by more sophisticated, integrative technological networks.

With the above examples, we have started to map the emergence of a potentially significant shift in urban strategies based around the idea of trying to shape how built spaces and electronically-mediated interactions work in parallel. Reviewing a set of innovative initiatives, it is apparent that planning initiatives are proliferating which try to actively shape the articulations between the development, and use of new media and communications technologies and urban places. Such initiatives are supported by widening efforts by planners, urban development agencies, transport bodies and media firms to understand the complexities through which electronically-mediated communications interact with land use, the urban realm, transport and face-to-face contact. Moreover, it is clear that the initiatives reviewed above are only the start. Virtually every western city worth it's salt now seems to uses 'cyber' and 'silicon' as obligatory prefixes in its urban marketing campaigns. Many beyond those discussed above are also exploring how they can address places and IT networks in parallel.

We would argue that the current growth of explicitly urban telecoms strategies and initiatives are broadly to be welcomed. This is for three broad reasons. First, they are acknowledgements that city-telecoms interactions are intrinsically bound up with contemporary metropolitan life. Second, these policies are based on much more sophisticated understandings of the complex and subtle relations between new media and urban life than those 'Death of Distance' of 'End of Cities" ideas generally implied by dominant 'information society' debates. And third, these proliferating policies suggest that the articulations between urban spaces and new media technologies are open to innovative, local, and planned interventions which can bring benefits which neither untrammelled market forces nor distant central state hierarchies can deliver.

But what might these nascent policies mean for urban planning and urban development more broadly ? Obviously, speculation is difficult in such an embryonic policy arena. In these conclusions we would therefore like only to address two key questions. First, are urban telecommunications initiatives likely to be able to succeed in shaping positive synergies between place-based interactions and development and electronically-mediated interactions and development? Or are they merely stylistic and symbolic, aimed add adding value and high-tech kudos to prestigious real estate developments? Second, what might these initiatives mean for broader notions of the 'city' and for ideas of integrated metropolitan-wide planning? More particularly, are these initiatives likely to reinforce and recreate new forms of socio-economic exclusion and environmental damage, or might they genuinely emerge as useful attempts to develop a more inclusionary and sustainable urban futures?

Turning to the first question, we must first sound some notes of caution. For, despite the widening range of initiatives, we remain highly cautious of their current usefulness, in terms of both their magnitude and direction. In terms of their magnitude, we must raise serious warnings against *overstating* the potential role of telecommunications and information technology in urban strategies. Whilst most approaches to urban strategies are still grappling with new ways of planning for transportation grids and urban places in parallel, it is still the case that electronic interconnections and networks are most often still hidden and taken-for-granted.

In terms of direction, it is clear that even when land use, transport and telecoms are considered in parallel, real progress will only come when two further problems are addressed. Firstly, policy makers will need to actively fight against prevailing assumptions (which are actually deeply embedded within western culture as a whole) that new technologies can somehow be rolled out as technical 'quick fix' solutions to complex urban problems. In a context where most urban policy-makers and planners lack knowledge and experience of the telecommunications sector, there is the danger that urban strategies could uncritically embrace the transformational rhetoric that characterises contemporary notions of technology 'impacts' upon the city that we touched on at the start of this paper. New telecommunications initiatives are still often intimately connected with utopian and deterministic ideas of technology's beneficial and liner impacts upon the social, environmental and spatial development of cities. Developing more nuanced and sophisticated concepts of the potential roles of telecommunications in urban strategy will require policy-makers to look more critically at the role of technology in contemporary urban strategies.

Achieving this, however, is difficult for another reason. Powerful media and technology firms are exploiting the hyperbolic rhetoric of 'cyberspace', the 'information superhighway' and the 'global information society' to enrol poorly-informed urban public policy makers into making local 'partnerships' to develop new information districts, communications corridors, and 'high tech' economic development zones of various sorts. Growing inter-urban competition, and the tendency for urban policy makers to jump on the latest policy band wagon, is being exploited by a wide range of consultants and media conglomerates. Such organisations are keen to add value and symbolic kudos to their own efforts to build up demand for new configurations of real estate and developed spaces, and all the associated technological hardware and services embedded in them. Public subsidies, discounted land deals, infrastructural assistance, credibility and the sheer marketing weight of public policy makers can do much to raise the profile of new, planned 'high-tech' spaces (and, therefore, developer profits).

But the real benefits of such initiatives to localities may be dubious or massively over blown because they remain inappropriate to real local needs Thus, planners and local policy makers need to educate themselves as quickly as possible about the burgeoning worlds of new media technologies,. They need to be wary of being seduced into expensive partnerships of dubious real local benefit by the lustrous promises of information age hyperbole. It is here that critical local debate about the real communicational needs of urban places, and the policy models that derive from these, is necessary.

Above all, planners must be sensitive to the important *symbolic* power of information and new media technologies, as signifiers of 'high tech' modernity. They must also be attentive to whose interests this symbolic power serves and how. Arguably, this symbolic power (which, by its very nature is very visible), is as significant, perhaps more significant in some cases, that real new applications or telecommunications infrastructures (which often remain unknown or hidden). Labeling a place 'cyber' this, 'silicon' that, or 'tele' something else, is an affirmation that it is switched in to global circuits of economic, cultural or social exchange via electronic networks. It is a potent, symbolic, attempt to lure in mobile capital, people and investment, which adds value to fixed infrastructure, land, and real estate even when they are clearly in tension with highly dynamic and mobile flows of services, media, information and money over telecommunications grids.

Such symbolism can take extreme forms. At one stage, for example, Edinburgh city council considered building an artificial satellite dish on its Maybury business park because BT argued that there was no technological need for there to be one in the City (which still has 'world-class' satellite facilities accessible from other places) (Graham, 1999). Of course, if local policy makers willingly embrace this symbolic power and use it creatively to their own advantage, through place marketing strategies and the like, all well and good. But they must be careful not to be duped by it, and not to uncritically believe all the promises of corporate and media firms that this new (publicly subsidies) media or IT network will miraculously solve all local problems of traffic congestion, social polarisation, economic development, environmental sustainability, and so on.

Which brings us on to our second question: what might these initiatives mean for broader notions of the 'city' and for ideas of integrated metropolitan-wide planning? Here, too, there is a need for concern. In the light of the above discussion, we clearly need to unpack the social assumptions and biases built into current urban telecommunications initiatives. We need to ask how urban telecommunications initiatives might link to wider urban debates around social equity, the public realm and culture, economic development, and environmental improvement. With such a strong supply push from powerful media and real estate interests, there are clearly dangers that urban telecommunications strategies are being configured in highly biased ways that might perpetuate and reinforce widespread existing trends towards social and spatial polarisation in urban areas.

The danger is that the foci of initiatives will centre overwhelmingly on configuring new media technologies according to the needs and geographies of affluent, privileged nodes, spaces and corridors in metropolitan regions whilst ignoring and excluding marginalised zones. The former, of course, are already at the vanguard of IT applications and are the 'hot spots' of demand for all forms of telecommunications and media applications and services (Graham and Marvin, 1996). In the context of liberalising telecommunications regimes, the risk is that market forces will encourage both corporate and media interests and urban policy makers to invest their efforts in communications corridors for the highly mobile, information districts for the information elites, and media consumption spaces for affluent professionals with high disposable incomes. In short, urban telecommunications strategies may simply work to extend the existing relational privileges of powerful zones, spaces and interests in the city.

Such fragmented policy 'packages,' superimposed as patchworks across the urban landscape, reflects wider trends within urban governance towards the collapse of the notion of coordinated, metropolitan-wide planning (Graham and Marvin, 1999). Complex patchworks of special interest zones and public-private governance initiatives are tending to replace systematic, metropolitan-wide public planning (Boyer, 1996). As Bosma and Hellinga (1998, 16) argue, in contemporary planning practices, "the primary matter of importance is no longer an integral approach, but the cheerful acceptance of regions as an archipelago of enclaves".

Urban telecommunications initiatives are clearly contributing to such an 'archipelago of enclaves'. The worrying thing about the urban telecommunications initiatives reviewed above may work to both reflect and sustain wider social polarisation trends. Thus, within cities, smart transport corridors and 'wired' enclaves might support forms of " telematics super-inclusion" (Thrift, 1996) for elite groups, allowing them to live in cocooned (often sometimes walled) enclosures, whilst still accessing personal and corporate transport and telematics networks. Meanwhile, however, a short distance away, in the interstitial urban zones, there are likely to be "off-line" spaces (Graham and Aurigi, 1997), or "lag-time places" (Boyer, 1996; 20). In these, often-forgotten places, access to the new technologies will remain highly problematic. Time and space will remain profoundly real, perhaps increasing, constraints on social life, because of welfare and labour market restructuring and the withdrawal of banking and public transport services.

But perhaps we are being too negative here. For we must also stress the positive potential of urban telecommunications initiatives as well as the need to be wary of their symbolic power and cautious of the possibilities that such initiatives may reinforce urban social and spatial polarisation. Progressive, inclusionary telecommunications and IT policies, integrated into particular urban strategies and designed to tie cities together rather than split them apart, might have important, positive roles in shaping the articulation between place-based and electronically-mediated realms. They might help, quite literally, to 'ground' the globally-integrating world of new media interactions, making them more meaningful in real places, real communities, real lives. Such initiatives may help to 'embed' new technological innovations in particular places, rather than just supporting an ever-more momentous de-localisation through market-driven forces of globalisation in the economy, society and culture. And, given enough stress on the needs of low income communities, such initiatives might actively counter wider urban polarisation trends (see Schön et al, 1997).

Indeed, one might even argue that *without* active, progressive resistance to untrammelled globalisation and the colonisation of local spaces by global media markets, mobilised through the rubric of creative, place-based, IT strategies, social need, the particularities of place, freedom of expression and local cultural diversity may tend to be squeezed out of the corporate, commodifying logic of globalisation. As Grossman (1995) puts it, "media conglomerates will not fill the vital educational, civic, and cultural needs" of real places and real cities. Strategies like the one in Manchester show that, just because privately-inspired urban IT initiatives have followed a particular trajectory, this does not mean that these cannot be challenged by incorporating wider social and environmental concerns into policy development.

Acknowledgements

We gratefully acknowledge the support of DEMOS and COMEDIA, particularly Ken Worpole, Bill Solsbury, and Liz Grenhalgh, which allowed us to undertake this research. Thanks also to the anonymous referees who provided very useful comments on earlier drafts. Responsibility for contents, of course, rests with the authors.

References

Association of County Councils (1996) *Green Communications: planning for telematics, teleworking and telecounties*, Environment Committee.

Bosma, K. and Hellinga, H. (1997), "Mastering the city". In Bosma, K. and Hellinga, H. (eds.) *Mastering the City*, Rotterdam; NAI Publishers.

Boyer, C. (1996) *Cybercities: Visual Perception in an Age of Electronic Communication,* New York: Princeton University Press.

British Telecom (1992) *A Study of the Environmental Impact of Teleworking*, A Report by BT Research Laboratories.

British Telecom (1997) *Telecommunications technologies and sustainable development - a guide for local Authorities.*

Carter, D. (1996), "'Digital democracy ' or 'information aristocracy '? Economic regeneration and the information economy". In B. Loader (ed) *The Governance of Cyberspace,* London: Routledge, 136-154.

Castells, M. (1997), *The Rise of the Networked Society*, Oxford: Blackwell.

CEC (1995) *Contributions of the information society to sustainable development,* ECOTEC (1996), *Information and communication technologies for sustainable technical development*, Final report.

CEC (1997) *European Spatial Development Perspective*, CEC: Brussels.

CEED (1992) The Environmental Impact of Teleworking, pp.10- 11, UK *CEED Bulletin* No 38, March-April.

Demos Collection (1997), *The Wealth and Poverty of Networks*, London: Demos.

Department of Energy (1994) *Energy, Emissions and the Social Consequences of Telecommuting*, Energy Efficiency in the US Economy, Technical Report One, DOE/PO-0026.

Economist, (1995), *The Death of Distance*, Telecommunications Survey, 30th September- October 6.

Environment (13th Nov 1997) Flexible working could reduce traffic.

Fitzpatrick, T. (1997), "A tale of tall cities", *The Guardian On-Line*, Thursday February 6, pp 9.

Gosling, P. (1997), *Government in the Digital Age*, London: Bowerdean Press.

Graham, S. (1998), "The end of geography or the explosion of space ? Conceptualising space, place and information technology", *Progress in Human Geography*, 22, 2., 165-185.

Graham, S. (1999), "Satellite dishes". n S. Pile and N. Thrift (eds) *City A-Z,* Wiley, London

Graham, S. and Aurigi, A. (1997), "Virtual cities, social polarisation and the crisis in urban public space", *Journal of Urban Technology* , 4(1), 19-52.

Graham, S. and Healey, P. (1998), "Relational theories of time and space: Issues for planning theory and practice", Paper submitted to *European Planning Studies.*

Graham, S. and Marvin, S. (1996), *Telecommunications and the City: Electronic Spaces, Urban Places,* London: Routledge.

Graham, S. and Marvin, S. (1999), *Splintering Networks/ Fragmenting Cities: Urban Infrastructure in a Global-Local Age,* Routledge: London.

Grossman, L. (1995), "Maintaining diversity in the electronic republic", *Technology Review,* November/December, 23-26.

Hill, D. (1997), *Cultural Industries in the Digital City*. Unpublished MA Dissertation, Manchester Metropolitan University.

Hill, S. (1988), *The Tragedy of Technology,* Pluto: London.

International Center for Communications (1997), *Building Smart Communities,* Guidebook, January.

Landry, C. and Bianchini, F. (1995), *Creative Cities,* London: Demos.

Lash, S. and Urry, J. (1994), *Economies of Signs and Space*, London, Sage.

McElvogue, L. (1997), "Bright sites, big city", *Guardian On-Line,* Thursday February 20, 2-3.

Martin, W. (1995), *The Global Information Society,* Aldershot, ASLIB.

Marvin (1997) "Environmental Flows,: Telecommunications and the Dematerialisation of Cities", *Futures,* Vol 29, No1.

Mitchell, W. (1994) "Building the bitsphere, or the kneebone's connected to the I-Bahn", *I.D. Magazine,* November 1994.

Mitchell, W. (1995), *City of Bits: Space, Place and the Infobahn* , Cambridge, Ma: MIT Press.

Mokhtarian, P. (1990) Relationships Between Telecommunications and Transportation, pp.231-242, *Transportation Research,* Vol.24A, No.3.

Montgomery, J. (1995) "Urban vitality and the culture of cities', *Planning Practice and Research*, 10,2: 101-109.

Moss, M. and Townshend, A. (1997), "Manhattan leads the Net nation". Available at http://www.nyu.edu/urban/ny_affairs/telecom.html.

Mulgan, P. (1991), "The changing shape of the city". In Stuart Hall and Martin Jacques (eds). *New Times*, London: Lawrence and Wishart.

Negroponte, N. (1995), *Being Digital,* London: Hodder and Stoughton.

Nilles, J., Carlson, F., Gray, P. and Hanneman, G. (1976), *The Telecommunications-Transport Trade Off,* Chichester; Wiley.

Office of Tecehnology Assessment (1995), *The Technological Resahping of Metropitan America*, Washington: Congress of the United States.

Roarke Associates, (1994) *Telecommuting offices - A proposal for congestion relief on London and S E England Rail services*, Roarke Associates.

Robins, K. (1995), "Cyberspace and the world we live in" in Mike Featherstone and Roger Burrows (eds.), *Cyberspace/Cyberbodies/Cyberpunk* London: Sage, 135-156.

Schön, D., Sanyal, B. and Mitchell, W. (eds) *High Technology and Low Income Communities,* MIT Press, Cambridge, Ma.

Schuler, D. (1996), *New Community Networks: Wired for Change*, New York: Addison Wesley.

Scottish Office (1993) *A national driver information and control strategy for Scotland*, A consultation document.

Siembab, W. (1992) *Metro Net, Fibre Optics and Metro Rail: Strategies for Development.*

Telecommunications for Clean Air (1994) South Coast Air Basin, *Institute for Local Self Government .*

Thrift, N. (1996)"New urban eras and old technological fears: reconfiguring the goodwill of electronic things", Urban Studies, 33(8), 1463-1493.

Thrift, N. and Olds, K. (1996) "Refiguring the economic in economic geography", *Progress in Human Geography*, 20(3), 311-337.

Times, (9 July 1997), "Teleworking could save a billion".

Wakeford, N. (1996), *Developing Community Intranets: Key Social Issues and Solutions*, Paper for BT.

Warren, R. (1989), "Telematics and urban life", *Journal of Urban Affairs,* 11(4), 339-346.

4 The Embeddedness of Digital Technologies: Any Implications for Cities?

Saskia Sassen

Digital networks are embedded in both the technical features and standards of the hardware and software, and in actual societal structures and power dynamics (Latour 1991; Lovink and Riemens 2002; Mackenzie and Wajcman 1999).[1] There is no purely digital economy and no completely virtual corporation or community. Power, contestation, inequality, hierarchy, inscribe electronic space and shape the production of software. Insofar as cities are one specific type of environment for multiple socio-political dynamics, one question is how do we go about studying embeddedness in this case? This essay examines some of the possible strategies that would allow us to do so. It is, then, not so much about the city as about studying the embeddedness of digital space.

Three analytic issues that capture various features of the embeddedness are the complex imbrications between digital and material conditions, the destabilizing of existing hierarchies of scale made possible by the new technologies which have the effect of repositioning cities in global networks, and, thirdly, the mediating cultures between these technologies and their users. These are the subjects developed in this paper after a brief discussion of the general features of embeddedness.

---

[1] Although using a different vocabulary, we can see Latour (1991) making a radical statement in this direction. Lovink and Riemens (2002) give us a detailed account of the multiple non-digital conditions (including neighborhood sub-cultures) that had to come together in order to create the enormously succesful city-wide digital internetwork called Digital City Amsterdam.

On Embeddedness of Electronic Space

The fact that electronic space is embedded and cannot be read as a purely technological condition, or merely in terms of its technical features, is illuminated by the nature of segmentations evident inside electronic space. One instance is captured in the differences between private and public-access digital networks.[2] The Internet is a different type of space from the private networks of the financial industry; and the firewalled corporate sites on the Web are different from the public-access portion of the Web. Financial markets, operating largely through private dedicated digital networks, are a good instance of private electronic space. The three properties of digital networks -- decentralized access, simultaneity and interconnectivity-- have produced strikingly different outcomes in the private digital space of global finance compared to the distributed power of the public-access portion of the Internet. Although the power of these financial electronic networks rests on a kind of distributed power, i.e. millions of investors and their millions of decisions, it ends up as concentrated power. The trajectory followed by what begins as the distributed power we associate with the public-access Internet, may assume many forms, and, in this case, one radically different from that of the Internet.

This difference points to the possibility that network power is not inherently distributive. Intervening mechanisms that may have little to do with the technology per se can re-shape its organization. To keep it as a form of distributed power requires that it be embedded in a particular kind of structure. We cannot take the distributed power and hence the democratizing potential of digital networks as an inevitable feature of this technology, as is so often the case in utopian readings, perhaps most prominently exemplified by John Perry Barlow's by now famous Declaration of Independence of Cyberspace (1997).

---

[2] To a large extent our thinking about electronic space and network power has been shaped by the properties of the Internet, disregarding the crucial differences between the public-access digital networks of the Internet and private digital networks to which there is no access no matter what one is willing to pay, e.g. private dedicated networks of financial services firms and wholesale financial markets

Beyond these issues of intentionality and use, lies the question of infrastructure and access (e.g., NTIA 1998; Petrazzini and Kibati 1999; Shade 1998; Sassen 1998: chapter 9; Thomas 1995). Electronic space is going to be far more present in highly industrialized countries than in the less developed world; and far more present for middle class households in developed countries than for poor households in those same countries (Jensen 1998; Harvey and Macnab 2000; Hoffman and Novak 1998). However, what needs emphasizing here is that there are very cheap ways of delivering access to the Internet, far cheaper than the standard telephone system, and hence that once such access is secured, the opportunities for low income households and communities, especially in the global south, can increase enormously (e.g., ITU 1999; Nadeau et al. 1998; Mele 1999).

Recognizing the embeddedness of electronic space, in my research I have come to regard the Internet as a space produced and marked through the software that shapes its use and the particular aspects of the hardware mobilized by the software (1999). These features can also function as an indicator of transformations in the articulations between electronic space and larger institutional orders. There are significant implications attached to the fact that one of the leading Internet software design focuses in the last few years has been on firewalled intranets for firms and encrypted tunnels for firm-to-firm transactions.[3] Both of these represent, in some sense, private appropriations of a "public" space.[4] Further, the growing interest in e-commerce has stimulated the development of software linked to identity verification, trademarks protection, and billing. The rapid growth of this type of software and its use in the Internet does not necessarily strengthen the public-ness of electronic space (e.g. Elkin-Koren 1996). This is espe-

---

[3] This saves companies the cost of private computer networks the associated staffing and servicing, and the cost of frame relay connections or the costs of using intermediaries for firm to firm transactions. Peer to peer

[4] An additional issue, one not examined here, is the privatisation of infrastructure that has also taken place since the mid-1990s. The backbone has been privatised where before it was financed by the US government, that is to say, taxpayers. This in turn changes the normative issues about private appropriations of Internet space as a public space. But it does so only partly, since it does not override the new distinction between privatised Internet space and public-access space, even if for a fee (for a resource to be public it need not necessarily be free). Internet space can remain public even if there is a fee to be paid for access, but privatised Internet space is not accessible no matter what the fee.

cially significant if there is less production of software aimed at strengthening the openness and decentralization of the Net as was the case in the earlier phases of the Internet. Far from strengthening the Internet's democratic potential as many liberal and neo-liberal commentators maintain, this type of commercialization can threaten it. It also carries major implications for the impact of democratizing initiatives.

However, electronic space remains a crucial force for new forms of civic participation, especially in its public-access portion. Non-commercial uses still dominate the Internet even though the race is on to invent ways of expanding electronic commerce and ensuring safety of payment transactions. But at the same time there has been a proliferation of non-commercial uses and users. Civil society is an energetic presence in electronic space. From struggles around human rights, the environment and workers strikes around the world to genuinely trivial pursuits, the Net has emerged as a powerful medium for non-elites to communicate, support each other's struggles and create the equivalent of insider groups at scales going from the local to the global (e.g., Perritt, Jr. 1996; Frederick 1993; Kobrin 1998; Ronfeldt et al. 1998; Rothkopf 1998).

Looking at electronic space as embedded allows us to go beyond the common duality between utopian and dystopian understandings of the Internet and electronic space generally. For instance, even as it reproduces masculine cultures and hierarchies of power, electronic space also enables women to engage in new forms of contestation and in proactive endeavors in multiple different realms, from political to economic. Further, in the context of globalization these initiatives can go global and bypass national states and major national economic actors, thereby opening a whole new terrain for initiatives of historically disadvantaged peoples and groups (e.g. Ronfeldt et al. 1998; Correll 1995; Mele 1999; Cleaver 1998).

Digital / Material Imbrications

Hypermobility or de-materialization are usually seen as mere functions of the new technologies. This understanding erases the fact that it takes multiple material conditions to achieve this outcome. Once we recognize that the hypermobility of the instrument, or the de-materialization of the actual piece of real estate, had to be *produced*, we introduce non-digital variables in our analysis of the digital. Obversely, much of what happens in electronic space is deeply inflected by the cultures, the material practices, the imaginaries, that take place outside electronic space. Much of what we think of when it comes to cyberspace would lack any meaning or referents if we were to exclude the world outside cyberspace. In brief, digital space and digitization are not exclusive conditions that stand outside the non-digital. Digital space is embedded in the larger societal, cultural, subjective, economic, imaginary structurations of lived experience and the systems within which we exist and operate.

For instance, producing capital mobility takes capital fixity: state of the art built-environments, well-housed talent, and conventional infrastructure --from highways to airports and railways. These are all partly place-bound conditions, even when the nature of their place-boundedness differs from what it may have been 100 years ago when place-boundedness was far more likely to be a form of immobility. Today it is a place-boundedness that is inflected or inscribed by the hypermobility of some of its components, products, and outcomes. Both capital fixity and mobility are located in a temporal frame where speed is ascendant and consequential. This type of capital fixity cannot be fully captured through a description confined to its material and locational features, i.e. through a topographical description (Sassen 2001: chapters 2 and 5).

Conceptualizing digitization along these lines allows us to recognize the ongoing importance of the material world even in the case of some of the most de-materialized digitized activities. This can be illustrated by the case of finance, one of the most digitized activities and one that involves a dematerialized instrument. Yet it cannot simply be thought of as exclusively digital. To have electronic financial markets and digitized financial instruments requires enormous amounts of materiel, not to mention human talent (which has its own type of physicality). This material includes conventional infrastructure, buildings, airports, and so on. Much of this material is inflected by the digital insofar as it is a function of financial markets. And much of the digital composition of financial markets is inflected by the agendas that drive global finance.

Digitization brings with it an amplification of those capacities that make possible the liquefying of what is not liquid. Thereby digitization raises the mobility of what we have customarily thought of as not mobile, or barely mobile. At its most extreme, this liquefying dematerializes its object. Once dematerialized, it gains hypermobility--instantaneous circulation through digital networks with global span. It is important --in my reading-- to underline that the hypermobility gained by an object through dematerialization is but one moment of a more complex condition. Representing such an object as hypermobile is, then, a partial representation since it includes only some of the components of that object, i.e. those that can be dematerialized. Much of what is liquefied and circulates in digital networks and is marked by hypermobility, remains physical in some of its components.[5] The real estate industry further illustrates some of these issues. Financial services firms have invented instruments that liquefy real estate, thereby facilitating investment and circulation of these instruments in global markets. Yet, part of what constitutes real estate remains very physical. At the same time, however, that which remains physical has been transformed by the fact that it is represented by highly liquid instruments that can circulate in global markets. It may look the same, it may involve the same bricks and mortar, it may be new or old, but it is a transformed entity.

We have difficulty capturing this multi-valence through our conventional categories: if it is physical, it *is physical*; and if it is liquid, it *is liquid*. In fact, the partial representation of real estate through liquid financial instruments produces a complex imbrication of the material and the de-materialized moments of that which we continue to call real estate. And so does the partial endogeneity of physical infrastructure in electronic financial markets.

---

[5] Much of my work on global cities (2001) has been an effort to conceptualize and document the fact that the global digital economy requires massive concentrations of material conditions in order to be what it is. Finance is an important intermediary in this regard: it represents a capability for liquefying various forms of non-liquid wealth and for raising the mobility (i.e. hypermobility) of that which is already liquid

The Destabilizing of Older Hierarchies of Scale

The complex imbrications between the digital (as well as the global) and the non-digital brings with it a destabilizing of older hierarchies of scale and often dramatic re-scalings. As the national scale loses significance along with the loss of key components of the state's formal authority, other scales gain strategic importance. Most especially among these are subnational scales such as the global city, and supranational scales such as global markets. Older hierarchies of scale dating from the period that saw the ascendance of the nation-state, continue to operate; they are typically organized in terms of institutional size and territorial scope: from the international, down to the national, the regional, the urban, to the local. But today's re-scaling dynamics cut across institutional size and across the institutional encasements of territory produced by the formation of national states (Sassen 2000b). This does not mean that the old hierarchies disappear, but rather that rescalings emerge alongside the old ones which can often trump the latter.

Existing theory is not enough to map today's multiplication of non-state actors and forms of cross-border cooperation and conflict, such as global business networks, NGOs, diasporas, global cities, trans-boundary public spheres, and the new cosmopolitanism. International Relations theory is the field which to date has had the most to say about cross-border relations. But current developments associated with various mixes of globalization and the new information and communications technologies point to the limits of IR theory and data. Its models and theories remain focused on the logic of relations between states and the scale of the state at a time when we see a proliferation of non-state actors, crossborder processes, and associated changes in the scope, exclusivity and competence of state authority over its territory, all partly enabled by these new technologies. Theoretical developments in other disciplines may prove important; especially relevant in the case of sociology's contribution is the type of network theory developed in economic sociology.

These transformations in the components of international relations and the destabilization of older hierarchies of scale can be captured in a variety of instances. For example, much of what we might still experience as the "local" (an office building or a house or an institution right there in our neighborhood or downtown) actually is something I would rather think of as a microenvironment with global span insofar as it is deeply internetworked. Such a microenvironment is in many senses a localized entity, but it is also part of global digital networks which give it immediate far-flung span. To continue to think of this as simply local is not very useful. More importantly, the juxtaposition between the condition of being a sited materiality and having global span, captures the imbrication of the digital and the non-digital and illustrates the inadequacy of a purely technological reading of the technical properties of digitization which would lead us to posit the neutralization of the place-boundedness of that which precisely makes possible the condition of being an entity with global span.

A second example is the bundle of conditions and dynamics that marks the model of the global city. Just to single out one key dynamic: the more globalized and digitized the operations of firms and markets the more their central management and coordination functions (and the requisite material structures) become strategic. It is precisely because of digitization that simultaneous worldwide dispersal of operations (whether factories, offices, or service outlets) and system integration can be achieved. And it is precisely this combination, that raises the importance of central functions. Global cities are, among other components, strategic sites for the combination of resources necessary for the production of these central functions. The cross-border network of global cities emerges as one of the key components in the architecture of "international relations."

Mediating Practices

There are multiple ways of examining the interactions between the new digital technologies and their users. There is a strong tendency in the literature to conceptualize the matter of use as an unmediated event, as unproblematized activity.

In contrast, a long-standing concern with what I have called "analytic borderlands" has led me to try to detect the mediations in the act of using the technologies. In my research I find that use is constructed or constituted in terms of specific cultures and practices through and within which users articulate the experience/utility of electronic space. Thus my concern here is not with the purely technical features of digital networks and what these might mean for users, nor is it simply with its impact on users. The concern is, rather, with this in-between zone that constructs the articulations of cyberspace and users.

This conceptualization clearly rests on the earlier proposition that electronic space is embedded and not a purely technological event. Thus electronic space is inflected by the values, cultures, power systems, and institutional orders within which it is embedded. If we were to explore these issues in terms of gendering, or specifically the condition of the female subject, we would then posit that insofar as these various realms are marked by gendering, this embeddedness of cyberspace is also gendered at least in some of its components, and, further, that so is cyberspace itself.[6] This is so even though there is enormous variabilty in this gendering by place, age, class, race, nationality, issue-orientation; at the same time, there are likely to be various situations, sites, individuals not marked by gendering, or marked by hybrid or queered genderings.[7]

The second consequence of this embeddedness is that the articulations between cyberspace and individuals -- whether as social, political, or economic actors -- are constituted in terms of mediating cultures; it is not simply a question of access and understanding how to use the hardware and the software. To some extent, these mediating cultures are likely to be shaped by gendering as well as other marking conditions.

***

---

[6] Much of what has been described for cyberspace in the specialized and general literature is explicitly or implicitly far more likely to be about particular groups of men because these have thus far dominated usage and produced many of the cybercultures (e.g. Holloway et al. 1999). Thus we also need more information about men who do not fit those particular groups.

[7] The concept of gendering has increasingly become problematic and is used here as shorthand for a complex bundle of issues. There is a vast critical literature on various aspects relating to gendering and feminist categories. For a broad range of issues see, for example, Wajcman 1991; Spelman 1988; Ong 1996; Reaume 1992. The notion of queering gender is, in this context, a powerful re-positioning.

The following two sections examine specific empirical instantiations of the major conceptual issues discussed in this first half of the article.

New Interactions Between Capital Fixity and Hypermobility

Information technologies have not eliminated the importance of massive concentrations of material resources but have, rather, reconfigured the interaction of capital fixity and hypermobility. The complex management of this interaction has given some cities a new competitive advantage at a time when the properties of the new ICTs could have been expected to eliminate the advantages of agglomeration, particularly for leading and globalized economic sectors. We have now come to understand that the vast new economic topography implemented through electronic space is one moment, one fragment, of an even vaster economic chain that is in good part embedded in non-electronic spaces. There is today no fully virtualized firm or economic sector. Even finance, the most digitized, dematerialized and globalized of all activities has a topography that weaves back and forth between actual and digital space. To different extents in different types of sectors and different types of firms, a firm's tasks now are distributed across these two kinds of spaces; further, the actual configurations are subject to considerable transformation as tasks are computerized or standardized, markets are further globalized, and so on.

To illustrate these issues I focus on three particular aspects of the interaction of capital mobility and fixity: the ongoing importance of social connectivity and central functions for global digitized economic sectors; the variety of locational options available to firms in partly digitized economic sectors; and the multiplication of the possible spatial correlates of centrality made possible by the new ICTs.[8]

a) The importance of social connectivity and central functions.

While the new technologies do indeed facilitate geographic dispersal of economic activities without losing system integration, they have also had the effect of strengthening the importance of central coordi-

[8] For a detailed presentation of the subject see Sassen 2001: chapters 2 and 5.

nation and control functions for firms and for markets. Major centers have massive concentrations of state of the art resources that allow them to "produce" the most strategic and complex of these central functions. It is not enough to have the infrastructure for ICTs. It also takes a mix of other resources: state of the art material and human resources, and the social networks that maximize connectivity. Much of the value added these technologies can produce for advanced service firms and advanced markets represents, then, a new type of urbanization economy insofar as achieving this value added depends on conditions external to the firms and markets themselves and to the technologies as such.

This new type of urbanization economy allows firms to maximize the benefits they can derive from the new technologies and to maximize their capabilities for operating globally. Even electronic markets rely on traders and banks which are located somewhere; for instance, Frankfurt's electronic futures market is actually embedded in a global network of financial centers, each of which concentrates resources that are necessary for Frankfurt's market to thrive.

A second fact that is emerging with greater clarity concerns the meaning of "information." There are two types of information that matter to advanced services firms. One is the datum, which may be complex but standardized and easily available to these firms: e.g. the details of a privatisation in a particular country. The second type of information is far more difficult to obtain because it is not standardized. It requires interpretation/evaluation/judgment. It entails negotiating a series of data sets and a series of interpretations in the hope of producing a higher order type of information. Access to the first kind of information is now global and immediate thanks to the digital revolution. But it is the second type of information that requires a complicated mixture of elements, not only technical but also social -- what we could think of as the social infrastructure for global technical connectivity. It is this type of social infrastructure which gives major financial centers a strategic role. In principle, the technical infrastructure for connectivity can be reproduced anywhere, but not the social connectivity (e.g. Meyer 2001; Garcia 2002).

When the more complex forms of information needed to execute major international deals cannot be gotten from existing data bases, no matter what one can pay, then one needs the social information loop and the associated de facto interpretations and inferences that come with bouncing off information among talented, informed people.[9] The process of making inferences/interpretations into "information" takes quite a mix of talents and resources.[10]

In brief, urban centers provide the mix of resources and the social connectivity which allow a firm or market to maximize the benefits of its technical connectivity.

b) Locational Patterns.

Information technologies have had a sharp effect on the spatial organization of economic activity. But this effect is not uniform: the locational options of firms vary considerably.

It is not simply a matter of reducing the weight of place. The scattered evidence for the last decade which saw the widespread use of information technologies by firms in a broad range of sectors allows us to identify three types of globally operating firms in terms of their locational patterns.

First, firms with highly standardized products/services see an increase in their locational options insofar as they can maintain system integration no matter where they are located. This might also hold for firms with specialized products/services that do not require elaborate contracting and subcontracting or suppliers networks, all conditions

---

[9] It is the importance for firms and markets of this complex type of "information" that has given a whole new importance to credit rating agencies, for instance. Part of the rating has to do with interpreting and inferring. When this interpreting is "authoritative" because originating with an established entity for its "production," it becomes "information" for the rest of us.

[10] Risk management, for example, which has become crucial with globalization due to the growing complexity and uncertainty that comes with operating in multiple countries and markets, requires enormous fine tuning of central operations. We now know that many, if not most, major trading losses over the last decade have involved human error or fraud. The quality of risk management will depend heavily on the top people in a firm rather than simply on technical conditions, such as electronic surveillance. Consolidating risk management operations in one site, usually a central one for the firm, is now seen generally as more effective. We have seen this in the case of several major banks: Chase and Morgan Stanley DeanWitter in the US, Deutsche Bank and Credit Suisse in Europe.

which tend to make an urban location more efficient. Data entry and simple manufacturing work can be moved to wherever labor and other costs might be lowest. Headquarters can move out of large cities and to suburban locations or small towns.

A second locational pattern is that represented by firms which are deeply involved in the global economy and hence have increasingly complex headquarter functions. Perhaps ironically, the complexity of headquarters functions is such that they get outsourced to highly specialized service firms. This frees up the headquarter to locate anywhere so long as they can access a highly specialized networked service sector somewhere, most likely in a city.

The third locational pattern is that evident in highly specialized networked service sectors. It is these sectors, rather than the headquarters, that benefit from spatial agglomeration at the point of production; in this regard, it is these firms rather than large corporate headquarters which are at the core of economic global city functions.[11] These firms are embedded in intense transactions with other such firms in kindred specializations and are subject to time pressures and the constraints of imperfect information discussed in the preceding section. Along with some of the features contributing to agglomeration advantages in financial services firms, this has the effect of rendering the network of specialized service firms more place-bound than the hyper-mobility of their products and of their professionals would indicate.

c) The spatialities of the center.

The combination of the new capabilities for mobility along with the advantages of urbanization economies for leading globalized economic sectors suggests that spatial concentration remains a key feature even in a global digital economy. But it is not simply a continuation of older patterns of spatial concentration. Today there is no longer a simple straightforward relation between centrality and such geographic

[11] These economic global city functions are to be distinguished from political global city functions, which might include the politics of contestation by formal and informal political actors enabled by these economic functions. This particular form of political global city functions is, then, in a dialectical relation (both enabled and in opposition) to the economic functions (see Sassen 1998: chapter one; Bartlett In Progress).

entities as the downtown, or the central business district. In the past, and up to quite recently in fact, centrality was synonymous with the downtown or the CBD. The new technologies and organizational forms have altered the spatial correlates of centrality.[12]

Given the differential impacts of the capabilities of the new information technologies on specific types of firms and of sectors of the economy, the spatial correlates of the "center" can assume several geographic forms, likely to be operating simultaneously at the macrolevel. Thus the center can be the CBD, as it still is largely for some of the leading sectors, notably finance, or an alternative form of CBD, such as Silicon Valley. Yet even as the CBD in major international business centers remains a strategic site for the leading industries, it is one profoundly reconfigured by technological and economic change (Fainstein 2001; Ciccolella and Mignaqui 2002; Graham and Aurigi 1997). Further, there are often sharp differences in the patterns assumed by this reconfiguring of the central city in different parts of the world (e.g. Marcuse and Van Kempen 2001).

Second, the center can extend into a metropolitan area in the form of a grid of nodes of intense business activity. One might ask whether a spatial organization characterized by dense strategic nodes spread over a broader region does in fact constitute a new form of organizing the territory of the "center," rather than, as in the more conventional view, an instance of suburbanization or geographic dispersal. Insofar as these various nodes are articulated through digital networks, they represent a new geographic correlate of the most advanced type of "center" (Veltz 1996; Yeung 2000). This is a partly deterritorialized space of centrality.[13]

---

[12] Several of the organizing hypotheses in the global city model concern the conditions for the continuity of centrality in advanced economic systems in the face of major new organizational forms and technologies that maximize the possibility for geographic dispersal. (See new Introduction in the updated edition of *The Global City* 2001). For a variety of perspectives see, e.g. Fainstein 2001; Orum and Chen 2002; Landrieu et al. 1998; Salomon 1996.

[13] This regional grid of nodes represents, in my analysis, a reconstitution of the concept of region. Further, it should not be confused with the suburbanization of economic activity. I conceive of it as a space of centrality partly located in older socio-economic geographies, such as that of the suburb or the larger metropolitan region, yet as distinct precisely becaue it is a space of centrality. Far from neutralizing geography the regional grid is likely to be embedded in conventional forms of communication infrastructure, notably rapid rail and highways connecting to airports. Ironically perhaps, conventional infrastructure is likely to maximize the economic benefits derived from telematics. I think this is an important issue that has

Third, we are seeing the formation of a transterritorial "center" constituted via intense economic transactions in the network of global cities. These transactions take place partly in digital space and partly through conventional transport and travel. The result is a multiplication of often highly specialized circuits connecting sets of cities. These networks of major international business centers constitute new geographies of centrality. The most powerful of these new geographies of centrality at the global level binds the major international financial and business centers: New York, London, Tokyo, Paris, Frankfurt, Zurich, Amsterdam, Los Angeles, Sydney, Hong Kong, among others. But this geography now also includes cities such as Bangkok, Seoul, Taipei, Sao Paulo, Mexico City. (See Yeung 2000; Sassen 2001). In the case of a complex landscape such as Europe's we see in fact several geographies of centrality, one global, others continental and regional.[14]

Fourth, new forms of centrality are being constituted in electronically generated spaces. For instance, strategic components of the financial industry operate in such spaces.

These various trends point to a profound transformation, but it is not one characterized by the neutralization of capital fixity, or of the built environment, or, in the end, the city. Rather than being neutralized, these emerge with renewed and strategic importance in *some* of their features, that is to say, not as a generalized condition but as a very specific condition. It is a particular type of built environment, of conventional communication system, of city, in brief, a particular type of spatiality that accomodates and furthers the new digital dynamics.

been lost somewhat in discussions about the neutralization of geography through telematics. For exceptions to this trend see Veltz 1996; Scott et al. 2001; Landrieu et al. 1998; Peraldi and Perrin 1996.

[14] Methodologically, I find it useful to unpack these inter-city transactions into the specific, often highly specialized circuits that connect particular sets of cities. For instance, when examining futures markets, the set of cities includes Sao Paulo and Kuala Lumpur. These two cities fall out of the picture when examing the gold market; this market, on the other hand, includes Johannesburg and Sydney (e.g. Harvey, In Progress).

A Politics of Places on Crossborder Circuits: Citizen Networks in a Global Digital Age

Digital networks are contributing to the production of counter-geographies of globalization. These can be constituted at multiple scales. Digital networks can be used by political activists for global or non-local transactions. But they can also be used for strengthening local communications and transactions inside a city. Recovering how the new digital technology can serve to support local initiatives and alliances across a city's neighborhoods is extremely important in an age where the notion of the local is often seen as losing ground to global dynamics and actors.

We can conceptualize these "alternative" networks as counter-geographies of globalization because they are deeply imbricated with some of the major dynamics constitutive of globalization yet are not part of the formal apparatus or of the objectives of this apparatus, notably the formation of global markets, the intensifying of transnational and trans-local networks, the development of communication technologies which easily escape conventional surveillance practices. The strengthening and, in some of these cases, the formation of new global circuits are embedded or made possible by the existence of a global economic system and its associated development of various institutional supports for cross-border money flows and markets. These counter-geographies are dynamic and changing in their locational features. And they include a very broad range of activities, including a proliferation of criminal activities.

Through the Internet local initiatives become part of a global network of activism without losing the focus on specific local struggles. (E.g., Cleaver 1998; Espinoza 1999; Ronfeldt et al. 1998; Mele 1999). It enables a new type of cross-border political activism, one centered in multiple localities yet intensely connected digitally. Activists can develop networks for circulating not only information (about environmental, housing, political issues etc.) but also political work and strategies.

There are many examples of such a new type of cross-border political work. For instance SPARC, started by and centered on women, began as an effort to organize slumdwellers in Bombay to get housing. Now it has a network of such groups throughout Asia, and some cities in Latin America and Africa. This is one of the key forms of critical politics that the Internet can make possible: A politics of the local with a big difference--these are localities that are connected with each other across a region, a country or the world. Because the network is global does not mean that it all has to happen at the global level.

The last few years mark a particular moment in the history of digital networks, one when powerful corporate actors and high performance networks are strengthening the role of private digital space and altering the structure of public-access digital space (Sassen 2000a). Digital space has emerged not simply as a means for communicating, but as a major new theater for capital accumulation and the operations of global capital. But civil society --in all its various incarnations-- is also an increasingly energetic presence in cyberspace. (For a variety of angles, see e.g. Rimmer and Morris-Suzuki 1999; Poster 1997; Frederick 1993; Miller and Slater 2000). The greater the diversity of cultures and groups the better for this larger political and civic potential of the Internet, and the more effective the resistance to the risk that the corporate world might set the standards. From struggles around human rights, the environment, and workers strikes around the world to genuinely trivial pursuits, the Internet has emerged as a powerful medium for non-elites to communicate, support each other's struggles and create the equivalent of insider groups at scales going from the local to the global of hierarchical power[15] (e.g. Calabrese and Burgelman 1999; see also Warf and Grimes 1997), rather than the space of unlimited freedom which is part of its romantic representation. The images we need to bring into this representation increasingly need to deal with contestation and resistance, rather than simply freedom and interconnectivity. The possibility of doing so transnationally at a time when a growing set of issues are seen as escaping the bounds of nation states makes this even more significant.

This is not the cosmopolitan route to the global. This is about the global as a multiplication of the local. These are types of sociability and struggle deeply embedded in people's actions and activities. They

---

[15] The Internet may continue to be a space for democratic practices, but it will be so partly as a form of resistance against overarching powers of the economy and

are also forms of institution-building work that can come from localities and networks of localities with limited resources and from informal social actors. We see here the potential transformation of women, "confined" to domestic roles, who can emerge as key actors in global networks without having to leave their work and roles in their communities. From being experienced as purely domestic, these "domestic" settings are tranformed into microenvironments located on global circuits. They do not have to become cosmopolitan in this process, they may well remain domestic in their orientation and remain engaged with their households and local community struggles, and yet they are participating in emergent global social circuits. A community of practice can emerge that creates multiple lateral, horizontal communications, collaborations, solidarities, supports. This can enable local political or non-political actors to enter into cross-border politics.

The architecture of digital networks, primed to span the world, can actually serve to intensify transactions among residents of a city or region, it can serve to make them aware of neighboring communities, gain an understanding of local issues that resonate positively or negatively with communities that are right there in the same city rather than with those that are at the other end of the world. Or it can serve to intensify transactions around the local issues of communities that *are* at opposite ends of the world. It is a peculiar mix of intense engagement with the local, with place, and an awareness of other "local" engagements across the globe. In brief, social activists can use digital networks for global or non-local transactions *and* they can use them for strengthening local communications and transactions inside a city or rural community.

Cyberspace is, perhaps ironically, a far more concrete space for social struggles than that of the national political system. It becomes a place where non-formal political actors can be part of the political scene in a way that is much more difficult in national institutional channels. Nationally politics needs to run through existing formal systems: whether the electoral political system or the judiciary (taking state agencies to court). Non-formal political actors are rendered invisible in the space of national politics. Cyberspace can accommodate a broad range of social struggles and facilitate the emergence of new types of political subjects that do not have to go through the formal political system.[16] Much of this becomes visible on the street. Much of urban politics is concrete, enacted by people rather than dependent on massive media technologies. Street level politics makes possible the formation of new types of political subjects that do not have to go through the formal political system in order to practice their politics. Individuals and groups which have historically been excluded from formal political systems and whose struggles can be partly enacted outside those systems, can find in cyberspace an enabling environment both for their emergence as non-formal political actors and for their struggles.

References

Adam, A. and E. Green. 1998. "Gender, Agency, Location and the New Information Society." Pp. 83-97 in B. Loader (ed.) *The Cyberspace Divide.* London: Routledge.

Bartlett, Anne. 2001. Politics Remade: Modernization and the New Political Culture in England." (Unpublished thesis. Department of Sociology, University of Chicago).

Bastani, S. 2000. "Muslim women on-line." *Arab World Geographer* 3, 1: 40-59.

Boneva B., R. Kraut, S. Kiesler and D. Frohlich. 2001. "Using the Internet for personal relationships: the difference gender makes." *American Behavioral Scientist* 45 (forthcoming).

Calabrese, Andrew and Jean-Claude Burgelman. 1999. Communication, Citizenship and Social Policy: Re-Thinking the Limits of the Welfare State. Rowman and Littlefield.

Cherny, L. and R. Weise (eds.). 1996. *Wired Women: Gender and New Realities in Cyberspace.* Seattle, Wa: Seal Press.

[16] I have made a parallel argument for the city, especially the global city, being a more concrete space for politics. In many ways, the politics of reivindication being enacted in cyberspace resonates with many of the activisms evident in large cities today: struggles against police brutality and gentrification, struggles for the rights of the homeless and immigrants, struggles for the rights of gays, lesbians and queers.

Cicollela, Pablo and Iliana Mignaqui. 2002. "The spatial reorganization of Buenos Aires." In Sassen (ed.) *Global Networks/Linked Cities*. New York and London: Routledge.

Cleaver, Harry. 1998. "The Zapatista Effect: The Internet and the Rise of an Alternative Political Fabric." *Journal of International Affairs* 51, 2: 621-640.

Correll, S. 1995. "The Ethnography of an Electronic Bar: The Lesbian Cafe." *Journal of Contemporary Ethnography* 24: 270-298.

Darkwa, O. and F. Mazibuko. 2000. "Creating virtual learning communities in Africa: challenges and prospects." *First Monday 5.5* (http://firstmonday.org/issues5_5/darkwa/index.html).

Elkin-Koren, N. 1996. "Public/Private and Copyright Reform in Cyberspace." *Journal of Computer Mediated Communication* 2, 2.

http://www.ascusc.org/jcmc/vol2/issue2/elkin.html

Espinoza, V. 1999. "Social networks among the poor: inequality and integration in a Latin American city." In B.Wellman (ed) *Networks in the global village*. Boulder, Co: Westview Press.

Fainstein, S. 2001. *The City Builders*. Lawrence, Kansas: Kansas University Press.

Featherstone, M. and R. Burrows. 1995. *Cyberspace/Cyberbodies/Cyberpunk: Cultures of Technological Embodiment*. London: Sage.

Frederick, Howard. 1993. "Computer Networks and the Emergence of Global Civil Society." Pp. 283-295 in Linda M. Harasim (ed.) *Global Networks: Computers and International Communications*. Cambridge, Mass.: MIT Press.

Garcia, Linda. 2002. "The Architecture of Global Networking Technologies." In S. Sassen (ed) *Global Networks/Linked Cities*. London and New York: Routledge.

Graham, S. and A. Aurigi. 1997. "Virtual Cities, Social Polarization, and the Crisis in Urban Public Space." *Journal of Urban Technology* 4, 1: 19-52.

Grint, K. and R. Gill. 1995. *The Gender-Technology Relation: Contemporary Theory and Research.* London: Taylor and Francis.

Hampton, K. and B. Wellman. 1999. "Netville online and offline." *American Behavioral Scientist* 43, 3: 475-92.

Hargittai, E. 1998. "Holes in the Net: The Internet and International Stratification." *INET'98 Conference: The Internet Summit* 21-24 July. Geneva, Switzerland.

Harvey, Rachel M. In Progress. Global Cities of Gold. (Dissertation Research, Department of Sociology, University of Chicago).

Harvey, A.S. and P.A. Macnab. 2000. "Who's Up? Global Interpersonal Temporal Accessibility." In D. Janelle and D. Hodge (eds.) *Information, Place and Cyberspace: Issues in Accessibility.* Amsterdam: Elsevier.

Haythornthwaite, C. and B. Wellman (eds) 2001. *The Internet in Everyday Life. Special Issue of American Behavioral Scientist.* (forthcoming).

Henshall, S. 2000. "The COMsumer Manifesto: empowering communities of consumers through the Internet." *First Monday 5.5* (http://firstmonday.org/issues5_5/henshall/index.html).Hoffman, D.L. and T.P. Novak. 1998. "Bridging the Racial Divide on the Internet." *Science* 280, 17 April: 390-391.

Holloway, S., G. Valentine and N. Bingham (Forthcoming). "Institutionalising Technologies: Masculinities, Femininities and the Heterosexual Economy of the IT Classroom." *Environment and Planning A.*

ITU (International Telecommunications Union). 1998. *Challenges to the Network: Internet for Development.* Geneva: ITU.

Jensen, M. 1998. *Internet Connectivity in Africa.* Report, January. http://demiurge.wn.apc.org/africa/

Knop, Karen. "Re/Statements: Feminism and State Sovereignty in International Law." *Transnational Law and Contemporary Probelms, vol 3, Fall 1993*: 293-344.

Kobrin, Stephen J. 1998. "The MAI and the Clash of Globalizations." *Foreign Policy* 112 (Fall): 97-109.

Landrieu, Josee, May, Nicole, Spector, Therese, and Veltz, Pierre (ed). *La Ville Eclatee.* La Tour d'Aigues: Editiones de l'Aube, 1998.

Latour, Bruno. 1991. "Technology is Society made durable." In J.Law (ed), A Sociology of Monsters. Lodnon: Routledge.

Loader, B. (ed). 1998. *Cyberspace divide: equality, agency and policy in the information age.* London: Routledge.

Lovink, Geert and Patrice Riemens. 2002. "Digital City Amsterdam: Local Uses of Global Networks." In Sassen (ed). *Global Networks/Linked Cities.* New York and London: Routledge.

Mackenzie, D. 1999. "Technological determinism." In W.H.Dutton (ed) *Society on the Line: Information Politics in the Digital Age.* Oxford: Oxford University Press.

Mackenzie, D. and J.Wajcman. 1985/1999. *The Social Shaping of Technology.* Milton, Keynes: Open University Press.

Marcelle, G. M. 1998. "Strategies for including a gender perspective in African information and communications technologies (ICTs) policy." Paper presented at the ECA International Conference on African Women and Economic Development, Addis Ababa, April.

Marcuse, Peter and Ronald van Kempen. 2000. *Globalizing Cities. A New Spatial Order.* Oxford: Blackwell.

Mele, C. 1999. "Cyberspace and Disadvantaged Communities: The Internet as a Tool for Collective Action." Pp. 264-289 in M.A. Smith and P. Kollock (eds.) *Communities in Cyberspace.* London: Routledge.

Miller, Daniel and Don Slater. 2000. *The Internet: An Ethnographic Approach.* Oxford: Berg.

Munker, Stefan and Alexander Roesler (eds). 1997. Mythos Internet. Frankfurt: Suhrkamp.

Nadeau, J., C. Lointier, R. Morin and M.A. Descoteaux. 1998. "Information Highways and the Francophone World: Current Situation and Strategies for the Future." *INET'98 Conference:*

The Internet Summit 21-24 July. Geneva, Switzerland. http://www.isoc.org/inet98/proceedings/5f/5f_3.htm

Nettime. 1997. Net Critique. Compiled by Geert Lovink and Pit Schultz. Berlin: Edition ID-Archiv.

NTIA (National Telecommunications and Information Administration). 1998. *Falling Through the Net II: New Data on the Digital Divide*. Washington, D.C.: NTIA July. USA. http://www.ntia.doc.gov/ntiahome/net2/falling.html

Ong, Aihwa. 1996. "Globalization and Women's Rights: The Asian Debate on Citizenship and Communitarianism."

Orum, Anthony and Xianming Chen. 2002. *Urban Places*. Malden, Ma: Blackwell.

Peraldi, Michel and Evelyne Perrin. (eds). 1996. *Reseaux Productifs et Territoires Urbains*. Toulouse: Presses Universitaires du Mirail.

Petrazzini, B. and M. Kibati. 1999. "The Internet in Developing Countries." *Communications of the ACM* 42, 6 (June): 31-36.

Poster, Mark. 1997. "Cyberdemocracy: Internet and the Public Sphere." Pp. 201-218 in D. Porter (ed.) *Internet Culture*. London: Routledge.

Rimmer, P.J. and T. Morris-Suzuki. 1999. "The Japanese Internet: Visionaries and Virtual Democracy." *Environment and Planning A* 31, 7: 1189-1206.

Reaume, Denise G. 1992. "The Social Construction of Women and the Possibility of Change: Unmodified Feminism Revisited." (Book Review). *Canadian Journal of Women and Law*, vol. 5, nr. 2, pp. 463-83.

Ronfeldt, David, John Arquilla, Graham Fuller, and Melissa Fuller. 1998. *The Zapatista "Social Netwar" in Mexico*. Santa Monica, Ca: RAND, MR-994-A.

Salomon, Ilan. 1996."Telecommunications, cities and technological opportunism." *The Annals of Regional Science*, 30: 75-90.

Sassen, Saskia. 2001. *The Global City: New York, London, Tokyo*. (New Updated Edition) Princeton University Press.

Sassen, Saskia. 2000a. "Digital Networks and the State: Some Governance Questions." *Theory Culture & Society Special Section on Globalization and Sovereignty*. Vol. 17, nr. 4 (August) pp.19-33.

Sassen, Saskia. 2000b. "Territory and Territoriality in the Global Economy." *International Sociology* vol. 15 (2): 372-393

Sassen, Saskia. 1998. *Globalization and its Discontents*. NY: New Press.

Schiffer Ramos, Sueli. 2002. "Sao Paulo: Articulating a cross-border regional economy." In Sassen (ed.) *Global Networks/Linked Cities*. New York and London: Routledge.

Shade, Leslie Regan. 1998. "A Gendered Perspective on Access to the Information Infrastructure." *The Information Society* 14: 33-44.

Shapiro, S. 1998. "Places and spaces: the historical interaction of technology, home, and privacy." *The Information Society* 14: 275-84.

Thomas, R. 1995. "Access and Inequality." In N. Heap, R. Thomas, G. Einon, R. Mason and H. MacKay (eds.) *Information Technology and Society: A Reader*. Milton Keynes: Open University Press.

Veltz, Pierre. *Mondialisation Villes et Territoires: L'Economie d'Archipel*. Paris: Presses Universitaires de France, 1996.

Wajcman, Judy. 1991. *Feminism confronts technology*. Cambridge, England: Polity press.

Wajcman, Judy. 2002. Information Technologies and the Social Sciences. Special Issue of Current Sociology.Vol. 50 (Summer)

World Information Order. 2002. World-Information Files. The Politics of the Info Sphere. Vienna: Institute for New Cutlture Technologies, and Berlin: Center for Civic Education.

Warf, B. and J. Grimes. 1997. "Counterhegemonic Discourses and the Internet." *The Geographical Review* 87: 259-274.

Yeung, Yue-man. 2000. *Globalization and Networked Societies*. University of Hawai'i Press.

5 The ICT Sector: Growth Engine for Urban Regions?

André van der Meer, Willem van Winden

Abstract

The ICT (information and communication technology) sector is at the centre of interest to cities, that see the sector as new source of wealth and employment. This paper describes and analyses ICT clusters and cluster policies in three European cities: Manchester, The Hague and Helsinki. The clusters show a very diverse development pattern, mainly due to differences in the spatial-economic structure of the cities and their position in their national urban systems. It is concluded that development chances depend on the initial size and quality of the cluster, the degree to which local knowledge-resources are exploited, the structure and quality of the local demand for ICT products and services, and very importantly, the quality of the urban living environment.

1. Introduction

ICT is the principal growth sector of the early 21th century. Its development is driven by accelerating technological evolution, resulting in a continuous stream of new products and applications on the market, eagerly used by companies and consumers. An important growth impulse for the ICT-sector is the liberalisation of telecom markets in Europe, which entails new players, lower prices, more competition and a much accelerated innovation process.

For cities, the growth of this sector opens perspectives for new economic development, wealth creation and employment growth. What are their chances to benefit, which factors are critical, and what should cities do to profit optimally from the sector's growth? These are the central questions of this paper.

The paper is organised as follows. In section 2, we start with a discussion of the definition and composition of the ICT sector, because there is much unclarity about this matter. Then, in section 3, we list the principal trends that affect the sector. A distinction is made between market trends, technological trends and institutional trends. Section 4 forms the core of the paper. Here, the question is addressed why and how urban regions would benefit from the information revolution. Section 5 gives an overview of possible urban policies to boost the ICT sector, enriched with many examples from our study in five European cities. Section 6 concludes, and derives strategic lessons for cities.

2. The ICT sector: definitions

Before discussing the impact of ICT on the urban economy, we should clarify what is meant by the ICT sector. Unfortunately, there is no generally accepted definition. Rather, a lot of distinctions and definitions co-exist. We will discus some of them.

The Dutch National Institute for Statistics (CBS) draws a distinction between ICT-manufacturing and ICT-services (CBS, 1999). ICT-manufacturing consists of all (sub-)sectors where goods are produced that contribute to the electronic infrastructure for both communication and the provision of information. Examples are producers of office machinery and computers; wires and cables; electronic components; telecommunication equipment. ICT services consist of all (sub-)sectors that supply services in the fields of telecommunication and information technology that make the utilisation of the (electronic) infrastructure possible (CBS, 1999). Both OECD (2000) and CBS (1999), place the ICT-manufacturers and ICT-services in a broader reality that is called the Information Economy or the Information Society. Besides the producers of the electronic infrastructure and the service providers that make use of the electronic infrastructure a third group of companies is distinguished: the suppliers of information products or "content". Examples are broadcasters, publishers, and advertisement agencies. There is not yet consensus about the exact definition of the "content"-sector. An important point of discussion is where to draw the lines within the wide range of entertainment, arts & culture related activities.

Other institutes use more refined distinctions, based on business processes. A recent example is a Booz, Allen & Hamilton (2000) study, in which the following subsectors are discerned:

- IT service providers and software developers. The software sector can be subdivided into packaged and embedded software. The former contains applications like word processing and games and application tools like spreadsheets and system infrastructure software. Embedded software is the software that is already incorporated in all kinds of devices like mobile phones and personal organisers, at the point of manufacturing.
- Providers of communication and network services, both fixed and mobile, which get the information to the end-user. Examples are telephone service providers.
- Producers of the public network equipment (transmission- and switching equipment, routers and cable producers): they produce the public electronic infrastructure. Examples are CISCO and Ericsson.
- Components and end-user equipment. Here we find the commodities that link the end users to the information society. The components sub-sector can be further divided into active components (like integrated circuits, tubes and LCDs) and passive components (like capacitors, connectors and resistors). Mobile and fixed private network equipment includes devices like telephone sets, mobile phones, television and radio transmission apparatus, switchboards and fax machines. IT hardware consists of office machinery, computers and other types of information processing equipment.

Some other institutes define a specific Internet and E-commerce sector. Banks and technology funds for instance discern infrastructure companies, Internet service providers, business to business E-commerce, business to consumer E-commerce, networking, security, investors, portals and advertisers (NRC, 2000).

In the following, we will take a broad perspective on the sector. We will include traditional sectors like telecom companies, equipment producers, as well as newer firms such as software developers, ICT consultants and Internet companies, but also electronic content providers. Additionally, we will pay generous attention to the "urban demand side" for ICT products and services.

3. Trends

The heterogeneity of the ICT sector makes it difficult if not impossible to list an exhaustive set of trends in "the ICT sector": each subsector is to a large extent subject to specific trends and developments. Nevertheless, some "fundamental" trends can be listed that affect virtually all segments of the sector. A distinction can be made between market trends, technological trends and institutional trends.

Market trends

- Growing market penetration of Internet. Internet use is expected to rise quickly, especially in Europe. That gives an enormous impulse for e-commerce activities; it is an incentive for even faster development of Internet applications, and other software products (interactive games, to mention only one), because the market grows by the minute.
- Exploding demand for telecommunications products and services. Drivers are the increased Internet penetration, increased volumes of data transport and the introduction of new services such as mobile Internet. New higher capacity infrastructure is developed to meet this demand.
- Falling prices in formerly regulated markets. Especially telecom services have become much cheaper as companies have to fight for market share, and will continue to fall.
- Ongoing mergers, acquisitions and strategic alliances. Some with the aim to exploit synergies, others to create scale. The time War-

ner-AOL merger in 1999 is an example of an effort to combine content production (Time Warner) and distribution (AOL). The merger trend leads to large companies, sometimes with substantial market power, but also contributes to the internationalisation of the ICT sector.

- Increased availability of venture capital. Finding risk capital used to be a difficult exercise, with banks reluctant to lend. Following the American trend, venture capital is now abundant in Europe. This enables high tech firms to engage in new projects, and spurs the number of start-ups. Figures released by the European Venture Capital Association reflect the explosion of activity in Europe's start-up scene: investments made in early-stage high-tech companies quadrupled in 1999 compared with 1998 to EURO 16 bn. (Financial Times, 7 June 2000).

Technological trends

- A fast pace of technological development. Technology developments are very rapid in all segments of the market. Therefore, ICT companies need to invest in R&D, or make sure to be at current with the latest trends. Time spans of investment projects are very short. Uncertainty rules.
- Electronic infrastructure. Many innovations increase the speed and capacity of existing data transportation (such as telephone lines), and new infrastructures are being developed, such as GPRS, UMTS and digital broadcasting technology. High-capacity infrastructure enables the development of all kinds of new services.
- The mobile revolution. Mobile telephones have become ubiquitous in a few years. In the next years, mobile Internet will probably take off. Furthermore, wireless applications can be used in many situations, for instance in cars (see box). Blue tooth technology enables computers to connect with each other, offering a range of new possibilities.
- Embedded technology. More and more, information technology is incorporated in conventional products. One example is the use of

telematics in cars: see box below. Mitchell (1999) provides a long list of other examples of "smart products"; to mention a few: clothes with build-in sensors that warn you when you need a shower; windows that can be opened/closed with your PC; intelligent fridges that automatically order new products when you take something out of it, etc.

Telematics in the car industry

Telematics has been a hot topic in the automotive industry, with most manufacturers seeing huge potential for offering value-added services – from e-mail access to navigational assistance - as wireless technology improves. Some estimates have suggested that half the new vehicles sold in North America, Japan and western Europe could have "embedded communications" modules by 2005. Ford Motor, the second largest car and truck-maker, plans to install telematics capabilities - for providing in-vehicle information, Internet and entertainment services - on virtually all its new vehicles by end-2004 and on about 3m by 2003. Ford is teaming up with Qualcomm, the US wireless chip company, to form a jointly owned company, called Wingcast, to spearhead its telematics initiatives.

Source: Financial times, July 31, 2000

Institutional trends

- Market liberalisation. Less than a decade ago, most European countries had only one supplier of telecom-services: the state monopolist. This landscape has changed drastically. Under strong influence of European policies most countries have freed their market to some extent; since then, many entrants have entered the various national markets. However, there is no "level playing field": each country has its own pace of liberalisation and many regulations are still of a national nature; some former state monopolists are in a more favoured position than others in their home market, which enhances their chances of expansion abroad. This hampers the sector's development, as scale economies can not be exploited on a European scale.

- De-regulation and re-regulation. During the last decade, in Europe a number of restrictions have been removed, which paved the way for a freer telecom market. New regulation has been developed to safeguard competition and prevent the formation of monopolies. However, there is still no European competition agency dedicated to the telecom sector.
- Intellectual property rights. The ease with which content can be downloaded and copied raises urgent questions on how to safeguard intellectual property rights. Recent trials in the US against Napster (a site that helped to access music files) show that the debate has only just begun.

4. The attractiveness of urban regions for ICT firms

The ICT sector is highly dynamic: institutions, markets and technologies are evolving rapidly. The central theme of this section is to what extent urban regions benefit from the explosive growth of the ICT sector. In the literature, there is support for the thesis that urban regions profit more than proportionally from the rise of ICT. Many authors (Abler, 1977; Alles and others, 1994; Graham & Marvin, 1996, Schmandt and others, 1990, Shields and others, 1993) find that both the development and application of ICTs (infrastructure, hardware and software) occur indeed in or near urbanised economic core regions. The best known examples are Silicon Valley in California, the M29 corridor near London, and Route-128 near Boston in the USA.

We suggest that ICT can be an engine behind urban economic growth in two ways. First, cities can benefit from the location of the new and fast growing ICT companies, that generate employment, added value and purchasing power. Second, cities may be the locations for emerging fruitful marriages between the ICT sector and "old" sectors, or, to put it differently, the place where "new combinations" emerge.

Why would urban regions, rather than other areas, benefit from the development of the ICT sector? In other words: what does the attractiveness of urban regions for ICT activity consist of? In line with our conceptual framework, we define attractiveness in terms of access to relevant provisions. A number of factors can be discerned. First, cities are nodes on (international) infrastructure networks, a very important amenity in a networked sector like the ICT sector. Second, cities provide access to concentrations of human resources, the most important resource for ICT firms. Third, cities facilitate access to markets and inputs: they form large concentrations of economic activity, which makes them attractive for ICT companies; and fourth, cities offer a quality of life that appeals to many ICT employees and firms. These arguments are elaborated below, enriched with examples from our case cities.

Figure 1

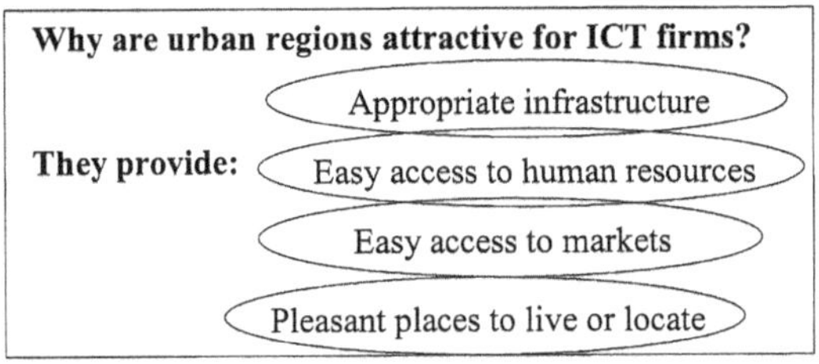

General access: cities as infrastructure nodes

The good accessibility of most urban regions contributes to their attractiveness as location for ICT companies. More than other sectors the ICT sector is a very flexible and "networked" sector: companies often co-operate with other (ICT) companies in projects, in ever-changing alliances. Accessibility is therefore critical: ICT companies often prefer locations near highways, railway stations and, if they operate internationally (many do), international airports. In Manchester, the southern fringe of the city, close to the airport, is a very popular spot for ICT companies and hosts an expanding cluster. E-commerce firms in The Hague (the headquarters of Amazon.com and Mapquest) greatly appreciate the nearness of Schiphol international airport.

The quality of the *electronic* infrastructure matters as well, in some instances. Because cities are such concentrations of economic activity, they are best provided with electronic infrastructure (see box), which makes them in turn attractive to ICT firms.

MCI WorldCom: principal cities the first to be connected by class fibre

The world-wide operating American telecom firm WorldCom (annual turnover: US$ 30 billion) is in the process of laying out a world-spanning glass-fibre network, which is to serve verbal and data traffic. The company addresses business users who have to process masses of voice, data, and Internet flows, such as banks, call centres and commercial offices. In the USA the network has by now branched out, interconnecting all the major towns where the customers are. By a transatlantic cable (laid out in a joint venture with Cables & Wireless; capacity: thirty gigabits a second), Europe was recently incorporated in that network. Amsterdam, London, Paris, Brussels, Düsseldorf and Frankfurt were the first European cities to be connected by a high-capacity fibre optic network, mainly because in those towns the principal customers are found. Local high-capacity networks have by now been constructed in more European towns, among which Stockholm, Zurich, Hamburg and Rotterdam.

source: Rotterdams Dagblad, 1998; MCI-WorldCom homepage

Some firms have special needs with respect to the quality of the information infrastructure; this particularly holds for companies that transmit enormous flows of data, such as Internet providers, electronic publishers or other media companies. Manchester can serve as example. Since 1998, TeleCity, the UK's only Internet Exchange outside of London, is located at the Manchester Science Park. This puts the area at the heart of UK Internet infrastructure development by offering the second major UK hub to the transcontinental Internet backbone (the other network access point in the UK is located in London). Businesses wanting to lease high speed Internet connectivity can now take advantage of cheaper rates as access is direct rather than via London. This complements the Manchester Network Exchange Point (MaNAP) set up by the University of Manchester in 1997 and together these services have been a key factor in the rapid growth of the Internet and e-business sector in Manchester. In Eindhoven, the university campus is equipped with broadband-infrastructure, and all students have a notebook that they can plug in anywhere. Several companies see this place as an opportunity to experiment with new web-based applications.

Access to human resources

Big cities are concentrations of "human capital", the most important resource for ICT companies. In many surveys, "availability of skilled staff" is mentioned as one of the primary location factors (see for instance Midas, 2000).

This point is illustrated in our case-study of The Hague: two new telecom service providers in the Dutch market decided to locate in The Hague, one of the reasons being that this city offered a large pool of specialised telecommunications specialists thanks to the presence of an incumbent firm (the former state monopolist) KPN.

Another asset of most cities is formed by its knowledge institutes, notably universities. They not only supply the city with trainees and new graduates, but in some cases also execute valuable basic research in information science that may be of importance to the local ICT industry. Perhaps even more importantly, universities are important breeding grounds for new ICT firms, started by entrepreneurial students or researchers who commercialise their ideas. In the ICT sector, business start-ups play an important role; many of the now-large ICT companies such as Amazon.com, were started from scratch by entrepreneurial youngsters in the 1990s. In the cities of Manchester and Eindhoven, this potential is clearly recognised. University based public-private companies offer space, legal and financial support to start-ups, and provide access to university facilities such as labs or powerful computers. In Eindhoven, the company takes a share in the start-up; in Manchester, start-ups have to pay royalties as a certain % of its turnover, after three years of profitable operation. In Manchester, in the last five years, 50 university start-ups have been supported, offering 200 jobs (Campus Ventures News, 1998). Some of them are expanding rapidly.

Access to demand for ICT products and services

The ICT sector is interwoven with other economic sectors: large parts of the ICT sector can be considered part of the business services sector. Cities are large concentrations of economic activity, which makes them attractive locations for ICT firms that prefer to locate near their clients. The most "heavy" users of ICT are corporate headquarters, financial institutions and the media industry (Saxenian, 1994; Hall, 1998). Particularly the rapid development of on-line banking and new media (online publishing, web-TV and radio) and entertainment services induce the ICT sector and the banking and media sectors to mix up. Financial centres are important ICT demanders: the financial sector has had the highest ICT expenses for several years now. This may explain the success of the cities of London, Amsterdam and New York in attracting ICT companies. The role of cities as concentrations of economic activity gives them a strong position to benefit from new combinations that arise from the application of ICT in existing sectors. Hall (1998) points at the fruitful marriages between ICT and typically urban sectors such as cultural, media and entertainment industries.

The integration of ICT with other sectors means that every city has chances to develop its ICT sector. It also suggests that each city develops a characteristic ICT-sector with its own "couleur locale", that fits the structure of the urban economy. The ICT-sector in Munich for instance is very strong in media applications, triggered by the high concentration of media companies in that city (Van den Berg, Braun and Van Winden, 1999). The ICT-sector in the famous Silicon Alley in New York is dominated by content creation, and addresses the concentration of publishers in the city. In Manchester, the ICT cluster is not that specialised, although the local financial institutes and media companies (ITV and BBC) are the principal demanders of ICT services. In Eindhoven, the ICT sector is more linked to the industrial sectors (that are strongly represented in the region). ICT companies develop applications for order processing, logistics applications or e-business solutions, and network-meetings are organised between old-economy (industrial) and new economy (ICT) companies, to discover new combinations and business opportunities.

In some cities, the ICT sector is less interwoven with the other sectors but plays a more independent role. The most outstanding example in Europe is probably Helsinki. The ICT cluster there is very large relative to other sectors, and very internationally oriented. Its development is driven by the global player Nokia, but also telecom service providers (such as Sonera) and small firms are increasingly international. Many firms are active in R&D.

Cities as pleasant places

Labour shortage and rapid business expansion are the causes of increasing global competition for ICT talent. From this perspective, the quality of life in the urban region –availability of comfortable housing, metropolitan ambiance, cultural conditions, green spaces, but also climate etc.- is an important condition to attract (or retain) staff to the city and thus for developing a cluster. Mitchell's (1999) point that in the end, the "pleasant places" on earth will be the main ICT centres might be somewhat extreme, but there is much truth in it. In the Netherlands, the explosive growth of Amsterdam's ICT cluster can be partly explained by its young, dynamic and cultural image. Its vivid urban life makes it -of course among many other factors- an attractive place for (young) ICT entrepreneurs.

Helsinki's cluster has been booming for several years now, but the limits to expansion are in sight. Up till now, the city's "knowledge base" has expanded thanks to the immigration of young professionals from other villages and cities in the country. However, it is not easy to attract foreigners to the remote city, with its cold and dark winters. The strategy of many rapidly expanding ICT companies is now to attract workers from the Baltic. Manchester has also problems in attracting –and retaining- its talent. Although the city is known for its nightlife and has a large (and expanding) cultural scene, it faces heavy competition for talent with London. The capital acts as a magnet upon talented people in any sector, also for ICT professionals. Many students who are educated at one of Manchester's universities leave the town for a job in London. In The Hague, which is part of the Randstad, the situation is a bit different. The city's ICT companies indicate no special problems in attracting ICT staff. This is not only due to quality of the city's environment, but also to the large labour market that is within reach: The Hague forms part of the Randstad area (with Amsterdam, Rotterdam and Utrecht), with five million inhabitants. A high percentage of the staff of The Hague based ICT companies don't live in the Hague or even its region.

Most cities' policymakers recognise the importance of quality of life as general condition for knowledge-intensive activities such as the ICT sector. Helsinki invests heavily in new cultural facilities (the new museum of contemporary art is an example), and upgrades the inner city area. The Hague is also upgrading and restyling its inner city, offering high quality housing and office space, and manages to attract highly educated people to the city. Manchester has a similar strategy: it seeks to improve the quality of its urban environment by restructuring the inner city and building expensive housing facilities, also in the city centre. The City of Eindhoven, although "leading in technology" in the Netherlands, has problems with its image as dull town. Some Eindhoven-based ICT companies try to benefit from Amsterdam's more dynamic image and present themselves as located "very near Amsterdam", to impress foreign partners or potential staff (indeed, it takes only 1.5 hours by train from Eindhoven to Amsterdam). The city council of Eindhoven fully recognises the city's image problem for several years and invests much in cultural facilities, shopping malls, etc, but an image does not change overnight.

5. ICT cluster policies in European cities

Almost every city wants to benefit from the growth of the ICT sector. All our case cities seek to be (or become) a principal centre of ICT, as can be read in the official strategy documents. The previous section has revealed general factors and circumstances that are conducive to ICT cluster development. The subject of this section is more policy-oriented. From our field work in five cities, we show examples of different types of ICT cluster policy, and critically asses them.

Evidently, a number of relevant variables that determine the development of a cluster cannot be influenced directly by urban policymakers but are rather national or European competencies. Examples are telecom market regulation, competition policy, tax policies, and research and education policies. Furthermore, the scope and room for urban policy varies widely among countries; for instance, Finland and The Netherlands are much more decentralised than France, which implies more degrees of freedom for urban government. The competencies regarding ICT cluster policies vary accordingly.

For the purpose of this study, it is useful to discern the following types of cluster-specific policies: infrastructure policies, ICT adoption policies, inward investment policies, creation of buildings dedicated to ICT companies, and finally, the development of experimental locations with broadband connections. See figure 2.

Figure 2: Policy instruments to promote the urban ICT cluster

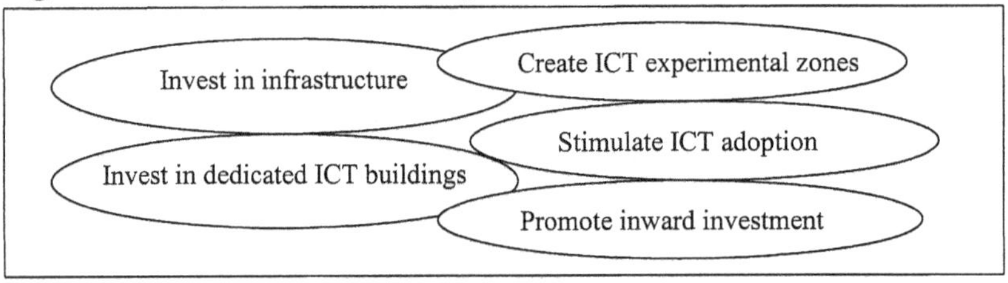

Here, we have deliberately left out the internal ICT policies of the cities (such as introducing local e-government, e-democracy and service provision), as they are of a different nature and serve different purposes

Infrastructure policies

Traditional telecom infrastructure provision (construction and exploitation) used to be a task of state-owned telecom monopolists, which owned and maintained the networks and also provided services. They provided national access without much regional discrimination. Additionally, in many countries local governments owned cable networks that transmitted television and radio programmes in the cities. In the last 10 years, in Europe the situation has changed (and is still changing) drastically. Most old telecom monopolists in Europe are privatised, have to compete with new carriers and allow service providers on their networks. New infrastructure-providers are developing electronic infrastructure (glass fibre networks), some Europe-wide (KPN Quest), some national, some global (MCI-Worldcom). These firms do tend to be spatially discriminating: generally, they follow demand, which means that big "telecom spending" agglomerations end up much better connected and served. Additionally, many of the local cable companies are "eaten up" by large telecom-providers. The former cable company of The Hague, Casema, for instance, is now owned by France Telecom. As a result, cities have lost control over their cable-networks. So, what are urban managers doing to influence the electronic infrastructure in their cities? We found several strategies.

- Connection to national/international backbones. The City of The Hague may serve as an example. This city is not a hub on the carriers' backbone, and is therefore less accessible than for instance Rotterdam and Amsterdam. To change this, the city seeks to convince carriers to extend their backbones to the city of Hague. It has commissioned an inquiry into the telecom-demand potential. The Hague proved to have potential comparable to Rotterdam but is not connected. The Hague also introduces carriers to potential clients in the region. The effectiveness of such policies can be questioned: whether carriers know the market less well than urban management remains questionable.

- Investment in broadband infrastructure in (parts of) the city as a means to speed up the development and adoption of new services, or help specific groups. The City of Eindhoven is creating an electronic “freeway”, free access to broadband, for some 80,000 of its inhabitants: this “knowledge neighbourhood” is to become a test-bed for new services. National government and companies are partners in the project. Helsinki has a similar project in the Arbianranta area.

- Creation of new infrastructure for intra-municipality data traffic. The city of Marseilles has infrastructure rolled out in the metro tunnels, connecting several local government agencies. A critical note could be why cities invest in dedicated infrastructure for the internal data traffic, instead of leasing lines from telecom operators whose core business is the provision of telecom infrastructure.

Dedicated buildings

Several European cities invest in dedicated buildings for ICT companies, in the hope that concentration of ICT firms in one building leads to synergies. A very special concept was developed in the inner city of Helsinki, the Lasipalatsi project. This is a public building that hosts the surprising combination of ICT firms, media firms, and leisure activities (a cinema and various bars). The centre not only offers office space for ICT and new media firms, it also serves as "showcase and disseminator" of new information technology developments. For instance, the bars and restaurants in the centre have touch-screens built-in in the tables, enabling visitors to order their drinks or to surf the net. Also, it brings producers and consumers of IT closer to each other: generally, developers of technology, the developers of its content and the everyday user are not in contact with each other (Lasipalatsin Mediakeskus Ltd., 1999). Lasipalatsi has a social goal as well: its open character should help to promote more equality in the information society by making IT easily accessible to all. The social function justifies public investment in the project. The building is owned by the city, but exploited by a separate public-private organisation that has substantial management freedom. This arms length construction works very well: it permits faster decision making. The Lasipalatsi contributes to the liveliness of the inner city of Helsinki, as it attracts many visitors.

Public support is not a necessary condition for creating dedicated ICT buildings: in The Hague, a private developer has turned an old switchboard-installation building (with the advantage of good external data connections) into an ICT centre. It hosts 33 small firms (total employment: 200), among which software developers, Internet designers, Internet access-providers, but also non-technical ICT related firms, such as lawyers specialised in intellectual property issues, a publisher of an IT magazine, and IT consultants. This developer has from the beginning aimed to realise synergies and co-operation in the building. Therefore, tenants are screened for their value added for the concept. A facilities provider in the building, Hobbit Facilities, plays a central "tying" role in the concept: to stimulate co-operation, it organises theme-sessions, sports events, and, for instance, an organised tour to CEBIT, the main computer fair in Europe. The concept seems to work well: there is a waiting lists for tenants, and the number of bankruptcies is very low compared to the national average. The role of the city of The Hague has been limited: it has only offered rent subsidy for firms during the first year the building is in operation.

Experimental areas or neighbourhoods

Some European cities explore the possibilities of the future by building futuristic urban quarters "loaded" with the latest technology. In that way, they try to make their city attractive as seedbed for new technological and infrastructure development, and to increase the number and quality of interactive services that are offered. Helsinki's Arabianranta area is an outstanding example, located at six km from the centre of Helsinki. The scheme's aim is to turn Arabianranta into one of the leading ICT-areas in Europe, where the functions of living, working, studying and recreation are mixed. The project is planned to be completed by 2014. The area already hosts 4,000 residents and 3,000 jobs; these numbers are intended to increase to 12,000 jobs and 8,000 residents respectively. An integral development strategy has been adopted, in which there is space for such complementary activities as research, business, education and housing and shopping. Already, 50 to 60 firms are present on the site, employing some 300 people. The number is growing for the site is very popular. Many of the small firms work for larger firms in the region, such as Ericsson, Nokia, Siemens and Sonera. The university of Art and Design is also located on the site, since 1985. Furthermore, new high quality housing and shopping facilities are being constructed. The electronic infrastructure of the Art and Design city is very sophisticated. A region-wide broadband fibre network is being implemented, with open access to all residents, schools and businesses. The intranet offers opportunities for teleworking, -shopping and –banking, tele-health and library services. The site proves attractive for large firms such as Sonera telephone corporation, and Nokia. For one thing, these firms can benefit from the superior quality of the electronic infrastructure by using the area as a test-site for new products, or as a showcase to display what is possible in the future.

Eindhoven is about to launch a similar project. Unlike Helsinki, the city will equip an existing urban quarter with broadband infrastructure. The quarter is very mixed, with high and low income residents, old and new houses, and stretches beyond the municipal borders. It will be an experimental zone for new services: not only will private companies offer services, but health care facilities and on-line municipal services will also be offered.

ICT adoption

Many cities run programmes to enhance ICT adoption by firms in their city. In this way, they not only seek to improve the competitiveness of "old economy" firms (mostly SMEs), but also indirectly stimulate the ICT sector by extending its client base. Manchester and The Hague run projects to help SME's make the shift to the information economy, by offering them free advice in the form of consultancy days (The Hague), or cheap advice (Manchester). Manchester uses ERDF money to fund these policies, and leaves the technical support to a university-based organisation; in The Hague, the project is carried out by Syntens, a regionally active organisation of the National Ministry of Economic Affairs. The impacts of these programmes on the development of the ICT clusters in the respective cities is unclear, although we feel that the effects should not be overestimated.

Another policy option is to stimulate ICT adoption by the population. In all our case cities, local government invests in computer skills of the population, by equipping schools and public buildings with computers, or by offering training. The use of new technologies can also be stimulated in a leisure-like setting: in Helsinki, visitors of a café or restaurant in the Lasipalatsi (Glass Palace) building are directly confronted with ICT. A touch-screen is integrated in the table, with which drinks can be ordered. In the end, the more eager urban population is to use new technologies and applications, the larger the market for local ICT entrepreneurs.

Inward investment

Many cities try to expand their local ICT sector with marketing efforts directed towards (foreign) ICT companies. In Manchester, the inward investment agency has designated ICT in general as spearhead sector. It offers US- and Canada-based firms general info on the region, and supports them to find a good location and staff, as well as tax reduction and subsidy facilities. In The Hague, acquisition efforts are tailored to telecom companies and e-commerce companies. The focus strategy pays: the city has managed to attract Amazon.com and Mapquest's European headquarters, and two large telecom providers. Marseilles wants to be an attractive location for telecom operators as well. To attract as many operators as possible, a one-stop shop agency has been created to welcome and receive new entrants. The city hosts already some 20 operators (Ville de Marseille, 2000). If cities manage to increase the number of operators, this may be beneficial for the citizens and companies when competition leads to more choice, better service and lower prices.

The regional scope of acquisition efforts varies among the cities. The inward investment agency of Helsinki, the Helsinki Metropolitan Development Corporation Oy HMDC, promotes international business not only in the city of Helsinki but in the entire region. HMDC is a joint-stock company owned for 52 % by the City of Helsinki. Other shareholders are the municipalities around Helsinki, the Uusimaa Regional Council and the Chambers of Commerce in Helsinki and Espoo. It markets the area as a suitable location for international business and co-operation and promotes it as a major business centre of the "new northern Europe". IT is one of its target sectors. Also, it helps small and medium-sized enterprises in the region expand their business abroad. In Manchester, the acquisition organisation works on a regional basis: it is owned not only by the City of Manchester, but also by the neighbouring cities of Salford, Trafford, and Tameside. It uses "Manchester" as brand name to sell the region. In the Hague Region, on the contrary, co-operation is not so smooth: the neighbouring cities of The Hague, Delft and Zoetermeer, have their own inward investment policies and strategies. The lack of co-ordination leads to unfruitful and inefficient inter-municipal competition. Distrust among the regional policy makers is the main cause.

6. Strategic lessons for cities

In this paper we have elaborated on the role of urban regions as attractive locations for ICT companies. Also, we described a number of policy options for cities to strengthen their position in this respect. What can cities learn from these analyses? A first, general lesson could be that it is wise to give room to private initiative. To create favourable conditions for highly dynamic ICT companies to flourish is of utmost importance. This includes provision of (electronic) infrastructure and appropriate business locations that suit today's demand, and avoidance of bureaucracy.

To speed up ICT business development, urban management can stimulate and/or enable experimentation with new technologies. Public investment in broadband infrastructure – for instance in experimental zones – might give a city a lead in ICT adoption by the population, but also speed up the development of useful services – if possible by local companies – as there it provides critical mass. This might give local companies a competitive lead in the future.

To arrive at effective policies, it is a prerequisite that urban managers should know the ICT sector very well. In too many cases, policies are ill-founded as urban officers have insufficient knowledge of what drives the market. One fruitful way to stay in touch with market developments in the ICT sector is to visit one of the many local "First Tuesday" events, meetings of Internet firms which are held in every large city.

Public-private co-operation is a prerequisite to arrive at effective and efficient cluster-policies. This holds in the marketing of the cluster, the attraction of new firms, in helping start-ups and in all other aspects of cluster policies. Public private co-operation helps to make optimum use of the knowledge and resources of the existing actors in the cluster. This again implies that officers involved in cluster policies need to be well educated and have sufficient "feeling" for the cluster.

Besides these general lessons, more specific ones can be derived.

- Starters policy. Create a starters-friendly environment: in the ICT sector, the starter of today can be the multinational of tomorrow.

Local government can act as provider of generic support, such as (cheap) accommodation and venture capital, but also as purchaser of the products of new companies, or as broker and connecting factor between starters and both the existing private sector and the knowledge institutions. Incubating is a lot cheaper – and often more effective – than attracting foreign business;

- Inward investment. If you want to attract high tech companies, define your target well. Rather than attract just any type of ICT-companies, make well-considered efforts to attract ICT-companies that fit the local economic structure or that fill gaps in the existing cluster. It is unfruitful to address the ICT sector as a whole, given the variety of firms' characteristics and needs within this sector. The concentration of The Hague on telecom-firms shows how effective a focus strategy can be. For inward investment agencies, it pays to use contacts and networks of existing firms in the regions with others. Involvement of existing private ICT firms in the inward investment stimulation schemes improves efficiency and effectiveness.
- ICT adoption. Stimulate adoption of ICT's by "old-economy" firms. Under some conditions, local government could encourage or speed up ICT adoption by lagging SMEs. Instruments are education schemes, demonstration projects, the provision of ICT consultancy days at reduced fees. Moreover, the city can organise working group sessions where ICT companies meet "old economy" firms in the region.

References

Abler, R. (1977), The telephone and the evolution of the American metropolitan system, in: De Sola Pool, I.(ed), *The social impact of telephone*, MIT press, London, pp 318-341

Alles, P., Esparza, A.; Lucas, S. (1994), Telecommunications and the large city – small city divide: evidence from Indiana cities, *Professional Geographer*, No. 46, pp 307 – 316

Berg, L. van den, E. Braun; W. van Winden (1999), Growth clusters in European cities: An integral approach, *Urban Studies* 38-1, 185-205.

Booz, Allen & Hamilton (2000), *The Competitiveness of Europe's ICT Markets*, Ministerie van Economische Zaken, Ministerial Conference, March 9-10, Noordwijk.

Campus Ventures News (1998), issue 4, *Statistics*, p 3

Centraal Bureau voor de Statistiek (CBS) (1999), *ICT-markt in Nederland* 1995-1998, Voorburg/Heerlen.

Financial Times (2000), European Venture Capital Report, June 7

Financial Times (2000), July 31, 2000, by Nikki Tait in Chicago and Tom Foremski in San Francisco

Graham, S.; Marvin, S. (1996), *Telecommunications and the city: Electronic spaces, urban places*, Routledge, New York

Hall, P. (1998), *Cities in Civilization*, Pantheon Books, New York

Lasipalatsin Mediakeskus Ltd. (1999), *Annual Report Urban Pilot Lasipalatsi 1999*

MIDAS (2000), *Environment for ICT*, internal publication

Mitchell, W.J. (1999), *E-Topia : Urban Life, Jim - But Not As We Know It*, MIT Press, Boston

OECD (2000), *Information Technology Outlook 2000*, Paris.

Ohmae, K. (1995), *The End of the Nation State; The Rise of Regional Economies*, The Free Press, New York

Porter, M. (1998), Clusters and New Economics, *Harvard Business Review* No 6 nov/dec 1998

Rotterdams Dagblad (1998), *Nieuw glasvezelnet koppelt Rotterdam aan wereldcentra*, 17-6-1998

Saxenian, A. (1994), *Regional Advantage: Culture and competition in Silicon Valley and Route 128*, Cambridge MA: Harvard University Press

Schmand, J., Williams, F., Wilson, R.; Strover, S.(eds) (1990), *The new urban infrastructure: Cities and telecommunications*, Praeger Publishers, New York

Shields, P. Dervin, B. Richter, C.; Soller, R. (1993), Who needs POTS-plus services? A comparison of residential user needs along the rural-urban continuum, *Telecommunications policy* , vol. 17, pp 563-587

Ville de Marseille (2000), *Marseille Economie, Bilan d'étape, Année 1999*, Direction Generale du Developpement Economique

5 Urban Authorities and the Harnessing of ICT

Bernd Fischer

1. The urban administrative actors

The findings presented in the following grew out of the research project "TeleCityVision" funded by the TSER-Programme of the European Commission. The project "TeleCityVision – Information Society and Urban Development in European Comparison" brought together seven research institutes from across Europe[1] and different scientific disciplines.[2] Goal of the empirical study was a better understanding of the processes prompted by the shift from an industrial to an informational society as the use and importance of information and communication technologies (ICTs)[3] increase for urban work and communication. In a socio-scientific perspective it was assumed that processes of

[1] The project partners were: Berlin Institute for Social Research, Germany; COMTEC at Dublin City University, Ireland; CTS at the Norwegian Technical University of Trondheim, Norway; ESI at the Free University of Amsterdam, Netherlands; THEMA at the University of Franche Comté, France; ETSIT at the Polytechnical University of Madrid, Spain; Department Geohumana at the Free University of Madrid, Spain; Interdisciplinary Centre for Comparative Research in Social Science, Austria/France; and ZTG at the Technical University of Berlin, Germany

[2] Sociology, political science, geography, economics, spatial planning

[3] The definition of information and communication technologies (ICTs) as used in the survey: "ICTs are a collection of technologies and applications which enable the processing, storing and transferring of information (hardware, software and infrastructure), provided by, and serving both the public at-large as well as the institutional and business sectors." This definition includes hardware (PCs, laptops, telephones, fax machines, mobile phones etc.), infrastructure (Fiber Optic Cable, Cable, ISDN, ASDL, ATM etc.), and software (world wide web, email, ftp etc.).

informatisation are not determined by the development of new technologies, but shaped and negotiated by social actors (compare Bijker et al. 1989; Graham/Marvin 1996; Woolgar 2002; for Germany Weingart 1989). Of particular interest to this study was thus *how* social actors learn to harness these new technologies.

The actors in urban development can be divided into three general categories: 1. public authorities, 2. businesses, and 3. private households. In our study, we focused on actors in public authorities, particularly the administrative staff within the executive and policy-makers within the legislative. With the ever-growing significance of ICT, many experts expect to see greater social differentiation, perhaps even a split within urban society into information-poor and information-rich (e.g. Castells 1997; Graham/Marvin 2001). City administrations and political decision-making bodies face a particular challenge to guarantee urban development that is fair to all social groups. The strategies developed by public authorities for dealing with ICT will have a decisive impact on the shape of cities in the information society. City administrations are not objects of a technology-determined development, but rather designers of this process – whether they decide to act or not.

According to Graham and Marvin, "Urban planners, policy-makers and governors are on the 'front line' in dealing with the effects of economic globalisation and restructuring; social and spatial polarisation and its associated knock-on effects on unemployment, poverty and crime; the crisis in urban environments and transport and infrastructure; and the physical restructuring of cities" (1996: 338). They go on to pose a crucial question: "How can those responsible for planning, managing and governing cities come to terms with threats and opportunities posed by telematics?" (Ebd.)

In this context, we need to keep in mind Ogburn's observation (1964) that the diffusion of technology is always accompanied by a cultural lag. Ogburn described how social change is driven by a tension between different cultural or social "variables," one more advanced, the other more delayed. In our case we are witnessing a tension between the technological development of a society and how public administration deals with these technologies in practice[4]. The lag suggests, as Fisher and Wright (2001) outline, "that the effects of a technology will not be apparent to social actors for some time after it is introduced to a society. As such, much of the discourse concerning the Internet [and ICT in general, B.F.] is ideologically charged, filled as much with the hopes and fears of individual authors as with the reality of the medium's effects". The notion of cultural lag can be misleading if we disregard the fact that there are various cultures or professional milieus, hence not only one but many cultural lags within any one administrative body or city. For this reason, we wanted to explore the differences between the "innovators" and "laggards" in the process of the diffusion of technology (Rogers 1983) and ICT practices within urban administrations. Cultural lags always imply that some time will pass until a technology is understood, harnessed, domesticated, or, in our case, urbanised. As we see it, the "cultural lag" is a kind of environmental pressure on the political-administrative system (Mayntz 1985: 35), which can be the basis for processes of learning, depending on how it is perceived by social actors. Due to different tasks and clients, this pressure can vary from department to department according to the specific socio-technical environment.

A key issue here involves the differences in ICT policies, applications and visions. How do such differences arise, and where do they lead? In order to answer such questions, we need to familiarise ourselves with the views and strategies of the actors, and how they learn to deal with ICT on an everyday basis.

2. Theoretical issues: How administrations learn

Because we identified a range of practices in dealing with ICT across the cities studied, we had to ask how administrations learn to deal with and adopt ICTs. How do structural preconditions influence the perception and agency of administrative and political actors? By viewing

---

[4] Many commentators have reproached Ogburn for a purportedly techno-deterministic view because he described technology as the main driving force in social change. In fact Ogburn stated that every "variable" can be the advanced one, but that a certain characteristic of modern times is that technological development is more advanced than other variables (Ogburn 1969: 88f).

the organisational learning process as a process of structuration, we could understand how the agency of the actors changes the structures in which they operate in a reflexive process (compare Giddens 1984, Ortmann 1997, Wilkesmann 1999, Rüegg-Stürm 2001). Orlikowski has shown how this "structuration" works by the introduction of new technologies in organisations (Orlikowski 1992). The organisations, even urban administrations, have to learn how to harness the new technologies towards the organisational objective.

Research on the sociology of organisations tells us that organisations do not learn as a whole. Only individuals within these organisations learn, if a given constellation allows learning. But the generation of new knowledge is relevant to the organisation only if it finds a place in the collective routine of the organisation (Wilkesmann 1999: 19f).

"Organizational learning occurs when individuals, acting from their images and maps, detect a match or mismatch of outcome to expectations which confirms or disconfirms organizational theory-in-use," write Argyris and Schön in a classic definition of organisational learning (1978: 19). Adds another scholar, "Organizations learn (...) by codifying individual insights, thus making them accessible to others for adoption, true adaptation, change, and generalized application" (Jelinek 1979: 157).

Figure 1: Complete cycle of choice (March/Olsen)

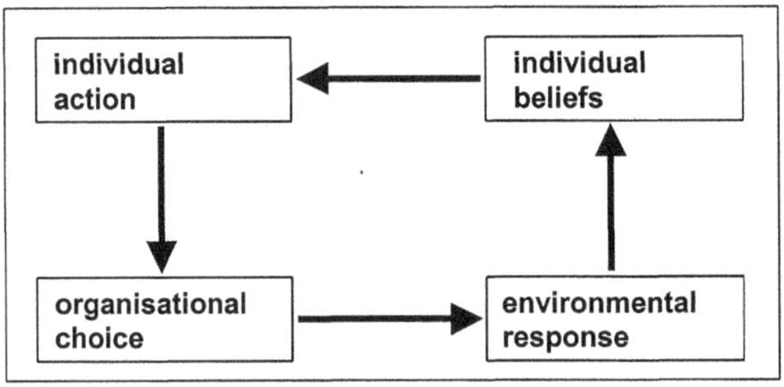

The "complete cycle of choice" is a model created by March and Olsen (1990) to show the interrelation of the individual and organisational level (Figure 1). Individual actions are transformed into organisational choice, which prompts a response from the environment. Individual actors interpret these responses against the background of their own beliefs and convictions; actors may change their actions according to their objectives.

March and Olson stress the importance of ambiguity in each step of the process. 1. Individuals commit varying amounts of attention to different situations, and different individuals commit varying amounts of attention to the same situation. 2. Organisations are social units, and not merely a tool with which to attain a particular economic goal. The decision-making process is tied to the definition and redefinition of "group interest". 3. Environmental acts are a result of the interplay of events, actors, and structures in the environment. Hence the same organisational act will have different responses at different times, and different organisational actions will have the same response. 4. Individual beliefs are constructed in an organisational setting. Organisational decision-makers do not have perfect information; beliefs and interpretations of the world are not tied directly to reality through experience.

The ability of an organisation to learn is dependent on a number of factors:

- if it is able to adopt individual knowledge,
- if the organisational choice is able to consider the environmental impacts,
- if individual actors are able to interpret the environmental reactions and to compare them to their own convictions and finally
- if they are able to implement these insights into their own agency.

In the following we focus on two levels. We first examine the "meso" level for the structural constellations that influence the perception and agency of urban administrative actors. Secondly, we study the "macro" level to see how the agency of the city administration, aggregated into ICT policies, might affect the environmental structure. In other words, we will try to analyse how the administrative actors become able to learn how to deal with and harness ICT, and how the implementation of new knowledge into policies can shape urban development.

The findings are based on the empirical work of the TCV research project. Based on a structural analysis of the European urban system, we chose differently structured cities in each partner country for case study (23 selected cities). Key to these case studies were qualitative interviews conducted with individuals working in or for the urban administration. After a phase of data-based theorising, the research team checked the hypotheses generated from the case studies in a quantitative investigation. The quantitative questionnaire survey covered 239 cities with a total of 1560 cases.[5]

Representative for the various actors involved as public authorities in urban development processes, the survey in both its qualitative and quantitative parts was directed at the departmental staffs for urban planning and economic development, as well as at communal politicians who were members of the committee for urban development or similar legislative organs.

3. Actors' perceptions and professional milieus

Within an actor's constellative preconditions of learning, we can distinguish individual, professional and institutional structures. As individual structures we consider the acceptance of technology and normative orientations. Professional structures consist of a certain milieu with special skills, values, self-definitions and modes of communication. The institutional constellation is the organisational structure with its division of labour, tasks, objectives and technological equipment. For an international comparison we also added cultural variables. The overall constellation influences the attitudes of individuals towards ICT.

Our data analyses showed that the most decisive structures are professional milieus and the institutional framework. Cultural and individual variables are relatively stable across the actors researched, with little variation even between genders.

ICT, together with its opportunities and risks, is perceived very differently. In order to understand the adoption of ICT by urban actors, it is

---

[5] Due to problems in finding city-types that work across Europe, the typology I use for possible future developments refers only to 795 cases in Germany.

crucial to know how actors perceive the technology. Lüdtke described the challenge as follows: "As seen by the actors, the functions of technology are diffuse in terms of the goals, use values, and secondary effects they pursue and experience, and thus compatible with differing and inconsistent structures of meaning" (Lüdtke 1994: 160). We tried to recognise and distinguish among such "differing and inconsistent structures of meaning" to explain divergent patterns of behaviour across a variety of urban administrative areas and cities.

In general, all the groups of individuals interviewed see the widespread application of ICT as an important factor of organisational modernisation and as an important agent of urban modernisation. In a more instrumental mode, there is no difference in how urban actors regard ICT. All actors emphasise that modernisation by ICT influences communication and work efficiency. However, they reject the idea that ICT effects socio-spatial patterns, or do not think that ICT is the primary cause of such changes. Nor do the respondents see any relation between ICT and environmental protection issues (see Burmeister/Hokkeler 1998). At the same time, a discussion of ICT and urban development does not seem to be a central issue in everyday (work-related) communication within administrations (Fischer/Helten 2002).

Furthermore, our investigations found that how people view and deal with ICT in practice is tied to the professional milieus within the departments in which they work. Actors negotiate the role and meaning of ICT at the level of departments – it is here that they apply ICT on an everyday basis. According to our data we can distinguish three professional milieus: urban planners, economic developers and politicians.

The institutional preconditions are, of course, different. Based on the division of labour within the administration, the departmental tasks shape the workplaces and equipment, and constrain the actors' scope for decision-making. In addition, some departments for economic development have been privatised, or experienced the out-sourcing of certain areas, which means that their organisational framework is not tied to the rest of the administration. This can be advantageous in terms of ICT equipment. Furthermore, politicians are not part of the administration, but they do interact with it closely in preparing and legitimating the administrative work and receiving information from it.

We found some indications that the respondents in urban planning departments are more involved than others in the ongoing discourse on ICT. Apart from their statements in the qualitative case studies, we recorded a relatively high response rate among urban planners. At the same time, they are quite sceptical or ambivalent in how they assess future urban concepts related to ICT. It seems that this combination of uncertainty and the need for help makes these actors more receptive of contributions from outside the administration.

The highest levels of ICT acceptance were found among politicians and economic experts. Both groups show a high level of equipment, modes of frequent use of ICT, and the most positive expectations towards increasing ICT use.

Figure 2: Different acceptance of ICT (N=1560)

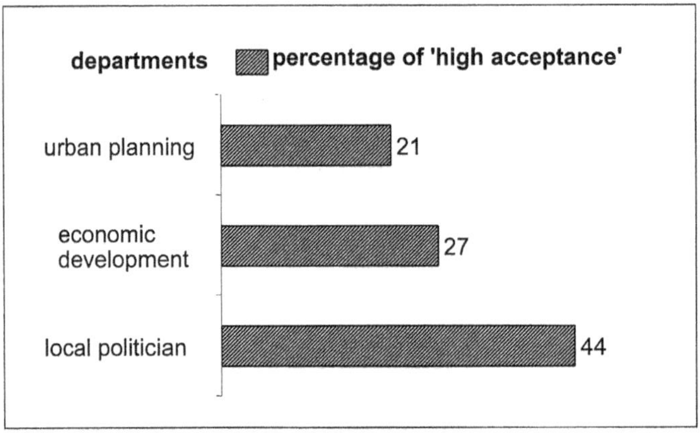

Indicator "acceptance" contains items of attitude and items of usage

While politicians clearly show the highest acceptance of ICT, economic developers are better equipped and have a higher level of everyday usage, thus the most routine and experience with ICT.

Figure 3: Differences in availability of PCs (N=1560)

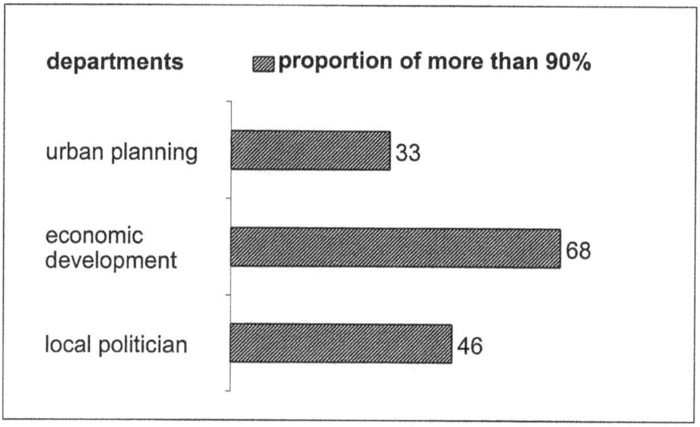

In the interviews, it also became clear that among economic developers e-mail and internet are everyday tools. Interestingly, though, they are valued in rather dry terms, despite the great importance accorded to them. For policy-makers, ICT tends to assume the character of a fetish, that is, they see ICTs as a catch-all solution for a range of problems, even though they might lack the relevant hands-on experience.

Figure 4: Unknown ICT strategy plans (N=1560)

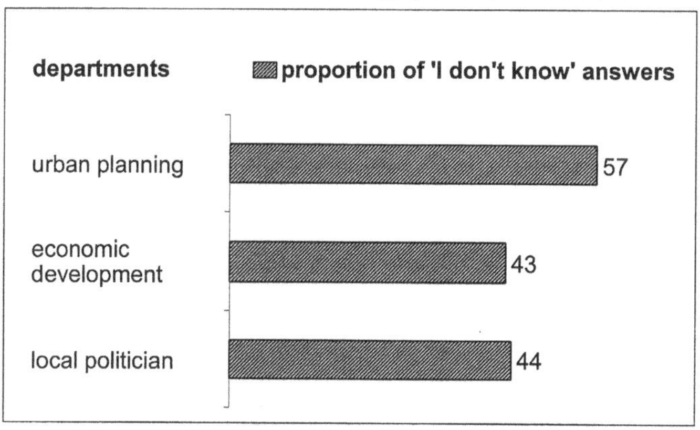

Despite this difference, both groups of interviewees display about the same amount of attention to and knowledge of urban ICT policy. In contrast, the most frequent response of urban planners, when asked about ICT strategy, was "I don't know". In other words, they are less informed, regardless of the existence of a common strategy plan or not.

The possible reasons for this variation are manifold: different occupational/professional roles, definitions and attitudes, corresponding to different duties and responsibilities, lead to a range of ICT strategies.

Urban planners, for example, are more oriented towards the local level because they feel in charge of equal access to urban infrastructure. They thus focus more on citizen participation. ICT is not seen as an appropriate instrument in this context, but rather as a tool for economic development. For their own work they accept ICT only as a drawing and writing tool (e.g. GIS, CAD).

Those involved in economic development think in more global terms and consider super/transregional processes. Their focus is on local companies, global competition and how to attract new business. ICT is broadly accepted as communication and information tool and also as appropriate help for economic development. Politicians, for their part, somehow range between these two categories.

Table 1 shows a simplified matrix of generalised statements from different categories of interviewees.

All interviewees assess the potential of ICT as a tool for economic development because it has the image of a contemporary means to build networks, to reinforce public relations, and to make cities more attractive to both prospective businesses and residents. ICT thus meets the objectives of economic developers but not of urban planners, who look to the local level and want to secure equal access to urban infrastructures for all urban citizens. ICT can, of course, be deployed for such purposes and some cities do so, but the average urban planner does not recognise it as such. ICT projects and the development of successful applications in the urban planning department largely depend on the commitment of single individuals.

Table 1: Differences among planners, economic developers and politicians

	Urban planners	Economic developers	Politicians
orientation	local	global	local and global
target group	citizens	companies	citizens as voters
objectives	equal access to urban infrastructure	economic success	depends on party platform and civil occupation of the respective politicians
image of ICT	tool for economic development	tool for economic development	tool for economic development
consequence	ICT does not meet objective	ICT does meet objective	it depends...

In line with such findings, we ascertained a range of attitudes towards ICT, ways of using and talking about ICT (e.g. "information flood" vs. "information offers"), different experiences, knowledge and expectations for the future – and thus a variety of strategies. Although occupational socialisation has become less relevant over the last decades, there still persist distinct professional milieus (Corsten 1999). These milieus have a specialised knowledge and a collective self-regulation system (Freidson 1986; Hudson/Sullivan 1990), and thus certain ideas as to if and how ICT can help them in solving specific problems or meeting objectives.

As far as expectations and assessments of ICTs go, urban planners tend to be more sceptical, economic developers to be more focussed on application, and politicians more enthusiastic.

Administrative employees and politicians differ in the types of action to take for future development. Politicians are much more open and positive towards new and ICT oriented practices and concepts. The average administrator is more cautious about new ideas or innovations, and tends to be guided by experience in everyday practice.

Many of the differences in the perception and use of ICT can be explained by the various motivations to employ ICTs. One of the main preconditions to learn in organisations is the motivation to change routine. Individuals will learn and change if they think innovations can solve certain problems. In our case, employees are willing to employ new technology only if they expect it will help them find solutions for perceived problems in their immediate work or in the city as a whole. We asked interviewees about the most important problems of their respective cities, and about the potential of ICT to solve these problems. In our example, urban planners see ICT as a "key instrument", but not as a solution to the problems they have identified. Politicians and economic developers, on the other hand, see opportunities for employing ICTs against certain problems. Both groups are motivated to learn.

In summary, we see that the more actors assume that ICT can help solve or ameliorate specific problems, the higher their motivation and interest to try out new applications. To urban planners, ICT does not address perceived problems, while to economic developers, it does.

ICT discourse within the various occupational and professional groups confirms the finding that better-equipped staffs are more inclined to see a broad field of possible applications. Lesser-equipped staffs are more sceptical towards ICT or, conversely, will develop a fetish about ICT as fix-all.

Due to the complex interplay of new technologies and learning, we see that innovations do not lead directly to appropriate solutions. But the opportunity and latitude to try out new applications can lead to more experience and greater possibilities for individual and organisational learning. As a result, actors will approach more appropriate or adequate strategies in dealing with ICT; innovative solutions might involve the rejection or reformulation of an existing routine, or even to an innovative strategy. Here, of course, there needs to be a minimum of organisational "slack" or "redundancy" to give to actors the opportunity to learn by playing with possibilities (Baecker 1994; Cyert/March 1995; Grabher 1994; Morgan 1993). Economic development departments thus share a more ICT friendly atmosphere because they see problem-solving possibilities and they have opportunities to try out new applications.

To conclude this section, we would like to present three factors we found that influence the ability to learn how to harness ICT in administrative organisations.

1. Professional milieus shape constellative structures in which individuals perceive ICT and its opportunities.
2. Individual motivations to learn result from the idea that changing routines can help solve or meliorate problems.
3. There exists a certain latitude for exploring new modes of agency for changing organisational routines.

In adapting the model of the complete cycle of choice, we have to take into account the structural framework in which individuals perceive the environment, take decisions and act. In the following figure, we adapted the boxes to the process observed in urban administrations and expanded the model with three catalysing functions in the arrows indicated by a, b and c.

Figure 5: Circle of harnessing ICT in organisations (adapted from March/Olsen)

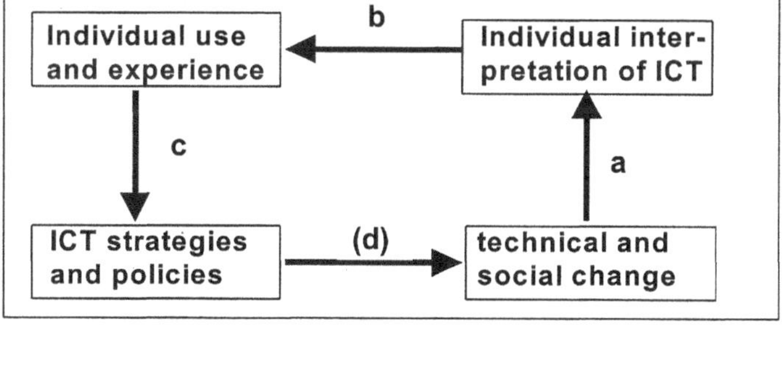

The added letters mean:

a) interpretation of ICT and its opportunities against the background of individual, professional and institutional constellations.

b) the individual expectation of ICT to solve specific problems and the willingness to change routines. Further, the institutional parameters and latitude for testing how ICTs could solve these problems.

c) The organisational adoption of changed individual practices into changed organisational routines.

The fourth arrow indicates an assumption that was not an actual area of our research:

d) new ICT strategies and policies influence technical and social change in the environment of the administrative body.

The three factors listed above – professional milieu, individual motivation, latitude – are essential for a reflexive learning process within departments and city administrations in finding appropriate solutions for employing ICT. Due to a myriad of problems and resources across departments and cities, it is problematic to adopt any one "best practice" model, as was often assumed by ICT experts and consultants in the past.

It would also be wrong to say that only the better equipped are able to learn and innovate. The actual process can be seen as a kind of gathering of knowledge from experience across contexts such as professional departments, whereby different applications of ICT help shape different ICT strategies and policies. It is impossible to predict what an appropriate strategy might look like.

What we can observe is the variation of these preconditions for organisational learning through individual agents among departments and cities, together with the various results of this reflexive process. The results are expressed in ICT strategies in different cities, which we can compare to the respective urban structure. This is the point where we turn from the agency of the actor to the development of the city.

4. Cities, ICT policies and increasing lags

Because organisational learning proceeds through individual interpretations and applications, ICT strategies vary both across departments and cities. Specific constellations, of course, also differ from city to city.

We can distinguish among the goals of ICT policies by the type of social relations they aim to promote. Objectives of ICT policies can be to improve relations between citizens and urban authorities, politics and urban authorities and business and urban authorities.

One common target in all urban ICT activities is to increase the economic attractiveness of the location. Providing excellent ICT infrastructure is a major concern. Another objective is to modernise the administration by making it more efficient. The main focus of local activities, however, is economic: to improve relations between business and urban authorities and to enhance the economic appeal of the location. Municipalities therefore support ICT projects that concentrate on infrastructure, information platforms, business-oriented services and the marketing of industrial spaces (compare Helten/Fischer 2002). As noted above, individuals regard ICT as tool for economic development. But cities vary greatly in the significance they accord to various ICT activities.

In the following section I will consider current and future developments of cities in relation to ICT activities. My remarks are mainly based on the German data, because European statistics did not provide the kind of in-depth knowledge of the cities and their actual development necessary to assess the validity of the findings.

For this purpose I synthesise existing typologies from Graham/Marvin (1996), Castells (1997), Sassen (1994), Krätke (1995) and Friedrichs (1997), as well as typologies that resulted from a structural analysis within the TeleCityVision project. I will propose a new typology as a basis for comparing ICT strategies, produced by cluster analysis combined with qualitative data. This enables us to distinguish five city-types in Germany.

1. European Multifunctional Cities (EMC): cities of European importance and with transnational control functions; significant, highly developed locations in the most important economic sectors.
2. National Multifunctional Cities (NMC): cities of national importance and supra-regional control functions; significant, highly developed locations in the most important economic sectors.
3. Rising Successful Cities (RSC): cities attractive to new enterprises, across all sectors; population structure is relatively young, image is positive.
4. Service and Information-processing Cities (SIC): cities whose economic potential is traditionally found in the tertiary sector; primarily production of immaterial goods; often administrative cities; unemployment relatively insignificant.
5. Industrial Cities (IDC): cities whose primary economic concentration has traditionally been in the secondary sector; unemployment is high, cities struggle with loss in standing and negative image; local industry can range from "old" coal and steel industry to "modern industries" in automobile and high-tech areas.

This typology does not claim to be exhaustive or of universal application. It is primarily meant to illustrate the possible different effects of various visions and strategies.

An example of this involves the different goals of ICT policies mentioned above. We can see that particularly European Multifunctional Cities (EMC), National Multifunctional Cities (NMC) and Rising Successful Cities (RSC) invest in an integrative ICT approach with a focus on economic development. Activities that address relations between citizens and administration lag behind.

The impact of different professional ICT knowledge is reflected in the respective ICT strategy. We applied a proposal from Korte/Grabow (1996) to distinguish among different urban ICT policies (comprehensive, focussed and uncoordinated strategies). It allows us to attribute certain strategies to each city type.

Figure 5: City types and ICT policies

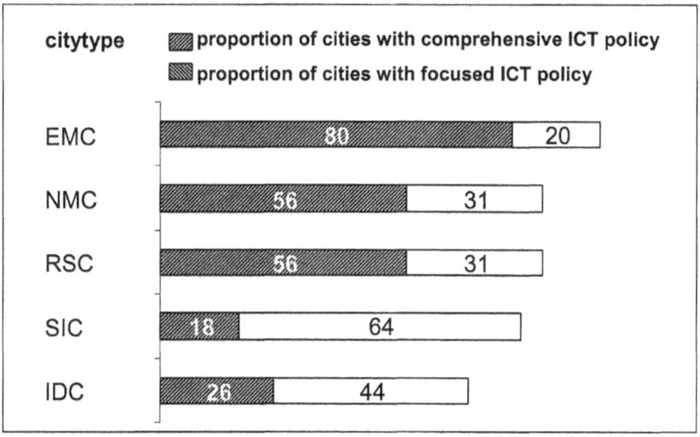

Actors in European Multifunctional Cities, National Multifunctional Cities and Rising Successful Cities rely mostly on comprehensive strategies, whereas Service Cities and Industrial Cities prefer a focussed strategy, which can mean concentrating on only one purpose of ICT application.

Figure 6: City types and ICT policies

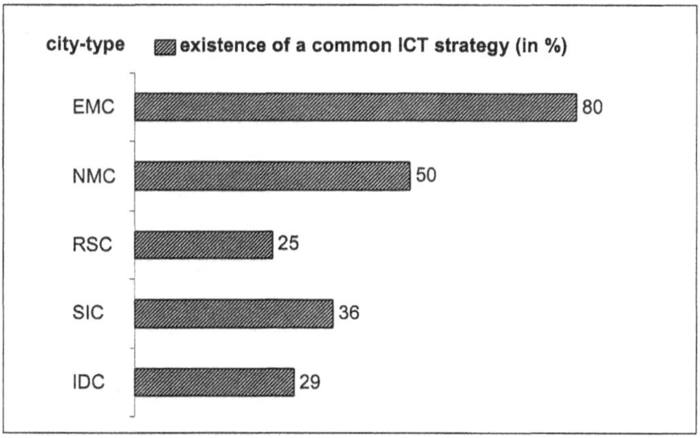

The existence of a common strategy in the municipality presents a similar picture. Here we see that Rising Successful Cites are less committed to formulating a common ICT strategy.

Further variables would yield similar findings. Generally, EMCs and NMCs are more active and better equipped, and their employees have a higher acceptance of ICT than employees in SICs and IDCs.

Here we need to recall the circle of harnessing ICT in organisations (relation d in Figure 5). Once a city commits to ICT – and the earlier it does –, its competitive advantage might very well lead to a further polarisation of the urban system. Actors in "important, successful and progressive" cities seem more able to use the advantages of ICT, while actors in cities burdened with substantial problems are less able to do so, even though ICT could serve as a means to balance out innovation deficits. This could result in increasing lags of innovation.

However, the fact that RSCs show only an average commitment to ICT policies, and promote only few ICT projects, raises the question as to why RSCs are so successful. To put it the other way around: does an ambitious ICT policy really promote the economic success of a city?

The following features are characteristic for RSCs: 1. average ICT policy: fewer activities than EMCs and NMCs, more activities than SICs and IDCs; 2. the municipality implements only a few ICT projects; 3. an above-average commitment in the area of planning and transportation; 4. participation in regional city-networks; and 5. actors have average expectations of ICT. Once ICT policies are found in RSCs, the concepts are mostly comprehensive and integrative.

In other words, the ICT complex alone does not seem to be the key to success. Other factors continue to determine the attractiveness of a city. For RSCs it is enough to act normally in various interrelated policy arenas in connection with city networks to attract entrepreneurs and (young) citizens. ICT policy seems to be a matter of adequacy rather than of quantity. We could compare these cities to economic development departments, where actors have ICT experience – they are neither euphoric nor sceptical, but motivated to explore opportunities to meet their objectives.

The more comprehensive commitment in EMCs and NMCs, and the more focussed commitment in SICs and IDCs, raises the question about the adequacy of such policies. We can assume that in cities of national and international importance, involvement in ICT policies suffices to maintain their multifunctional and central role. SICs and IDCs might run the risk of insufficient innovation. That said, we have to keep in mind that an ambitious ICT policy alone cannot solve such problems as unemployment, decline in the industrial sector, ageing of the population, or negative image.

Although generally accorded secondary importance, improving communication between citizens and the municipality could offer a promising approach to stable cities with problematic economic and social structures. We agree with the notion of Graham/Marvin that "attention needs to focus on the social organisation of telematics systems" (Graham/Marvin 1996; 343). In our study, the respondents also tend not to consider the effects of ICT on participation, democracy, ecology and transportation, communication and interactivity. Precisely the promotion of such "soft" location factors could prove a special achievement of new technology and its usage in cities.

It is also astonishing – especially because of the economic potential involved – how seldom public-private partnerships play a role in the implementation and funding of ICT projects, although nearly all groups of interviewees stated that private companies and investors are the most important actors in the process of informatisation.

Finally, local actors attribute remarkably little importance to the level of national government for financial support of local ICT projects or as actor in the informatisation process itself. In the past we witnessed the success story of the state-backed French Minitel campaign and the French government's support of telecommunications infrastructure (Graham/Marvin 1996; Castells 2001). Graham/Marvin (1996; 345) assume that such efforts would be an important condition for a future-oriented ICT policy, especially in light of globalisation, where importance shifts from the national to global and local levels.

The efforts of the European Commission in promoting ICT seem to have unintended consequences. Projects in more successful cities of international importance have, of course, received funding, while peripheral and problematic cities tend to be neglected. Should that impression be confirmed in further investigations, we would need to examine whether the European Commission might not be contributing to the polarisation of the urban system in its funding policies. In the meantime, the e-inclusion issue has become an important aim of the Commission's activities.

5. Resulting policy implications

We started by stating that the ICT complex is seen as a driving force in the modernisation and transformation process of cities. The process of modernisation, however, is mainly reduced to its economic aspects. Astonishingly enough the social, cultural and spatial implications of ICT are neglected. Such broader implications are not a central concern of public authorities, who restrict their expectations of ICT to a way of improving communication among departments and facilitating access to administrative services. Here we see that the shaping and use of ICT are related to a range of professional experiences, views and strategies.

One of our goals was to find out how to distinguish structures that promote innovative actions from structures that hinder innovation. Professional milieus create and allow certain ways of learning and experiencing. The results of that process are various perceptions and strategies to use and apply ICT. Because it is impossible to make prior judgements about the adequacy of ICT policies, each municipality depends on a development strategy that takes into account various experiences and objectives.

Cities run the risk of focussing too much on only one aspect of ICT – mostly economic development – and lack opportunities to explore the full spectrum of ICT. Together with the issue of cultural lag, this means that perceptions of ICT in different cultural milieus or structural city types can become one-dimensional or even ideological. Applications that focus on societal, cultural or participatory issues are neglected or overshadowed by issues believed to be more important. In addition to the knowledge deficit resulting from cultural lags, the ongoing process of functional differentiation and professional specialisation makes it even more difficult to communicate across administrative boundaries. This effect is amplified by the organisational structure of administrative organisations, which literally oppose network structures usually supported by ICTs.

I would like to conclude with several proposals for improving the future quality of urban governance, to help prevent a digital divide among both professions and European cities:

First, we need to understand that the shaping of technological systems as ICT is an ongoing social process of learning and negotiating. It is necessary to explore and research the potential of new technologies in order to know about possible use, applications and consequences.

Finding adequate strategies is dependent on the actor's ability to learn and adjust routines, and to share and exchange experience. This is also the reason why ICT departments in municipalities should include representatives from different departments, and not only ICT experts, to consider different experiences, aims, and professional orientations.

To supplement EDP training, education measures should include strategies, objectives and effects involved with the application of ICT.

Finally, national and European ICT policies need to consider support of local ICT policies as a supra-local task. The transfer from more advanced to less developed cities is necessary to prevent further polarisation and cultural lags.

Goal of all these activities is not the implementation of a universal best-practice model. Municipal administrations need to be enabled to develop and tailor their own specific strategies for harnessing ICT to their particular challenges.

References

Argyris, Chris; Schön, Donald A. (1978): Organizational learning: A theory of action perspective. New York: MacGraw-Hill

Bijker, Wiebe E.; Hughes, Thomas P.; Pinch, Trevor F. (Eds.) (1989): The Social Construction of Technological Systems. Cambridge Mass.: MIT Press

Burmeister, Klaus; Hokkeler, Michael (1998): Nachhaltige Stadtentwicklung in der Informationsgesellschaft? In: Bundesamt für Bauwesen und Raumordnung: Informationen zur Raumentwicklung, Volume 1, 1998, pp. 31-40

Castells, Manuel (1997): The Rise of the Network Society. The Information Age: Economy, Society and Culture. Volume 1. Malden Mass., Oxford: Blackwell

Corsten, Michael (1999): Treulose Arbeitsindividuen ohne berufliche Bindung – Mythen und Anti-Mythen zur Krise beruflicher Sozialisation. In: Honnegger, Claudia; Hradil, Stefan; Traxler, Franz (Eds.): Grenzenlose Gesellschaft? Teil 1. Opladen: Leske und Budrich, pp. 290-306

Fischer, Bernd; Helten, Frank (2001): TeleCityVision. Perception, Acceptance and Strategies of Urban Administrative Actors. Results of the quantitative survey. Berlin: BIS Research report

Fisher, Dana. R.; Wright, Larry Michael (2001): On Utopias and Dystopias: Towards an Understanding of the Discourse Surrounding the Internet. Journal of Computer-Mediated Communication 6 (2). (http://www.ascusc.org/jcmc/vol6/issue2/fisher.html)

Freidson, Eliot (1986): Professional Powers: A study in the institutionalization of formal knowledge. Chicago: University of Chicago Press

Friedrichs, Jürgen (1997): Eine Typologie westdeutscher Städte und Muster ihrer Entwicklungen 1970 bis 1990. In: Friedrichs, Jürgen (Ed.): Die Städte in den 90er Jahren. Opladen, Wiesbaden: Westdeutscher Verlag, pp. 67-90

Grabher, Gernot (1994): Lob der Verschwendung. Redundanz in der Regional-entwicklung: Ein sozioökonomisches Plädoyer. Berlin: Edition Sigma

Grabow, Busso; Werner B. Korte (1996): Telematik, Teledienstleistungen und Kommunalpolitik. Difu Aktuelle Informationen. Berlin

Graham, Stephen; Marvin, Simon (1996): Telecommunications and the City. Electronic Spaces, Urban Places. London, New York: Routledge

Graham, Stephen; Marvin, Simon (2001): Splintering Urbanism. Networked Infrastructures, Technological Mobilities and the Urban Condition. London, New York: Routledge

Helten, Frank; Fischer, Bernd (2002): Stadtraum und Informatisierung. Sozialer und technischer Wandel im Stadtentwicklungsprozess. In: Berking, Helmuth; Farber, Richard (Eds.): Städte im Globalisierungsdiskurs. Würzburg, pp. 139-168

Hudson, Randy; Sullivan, Teresa A. (1990): The Social Organisation of Work. Belmont, CA: Wadsworth

Jelinek, Mariann (1979): Institutionalizing innovation. A Study of Organiza-tional Learning Systems. New York: Praeger

Krätke, Stefan (1995): Stadt, Raum, Ökonomie. Eine Einführung in aktuelle Problemfelder der Stadtökonomie und Wirtschaftsgeographie. Basel, Boston, Berlin: Birkhäuser

Lüdtke, Hartmut (1994): Alltagstechnik im Kontext von Lebensstilen. In: Noller, Peter; Prigge, Walter; Ronneberger, Klaus (Eds.): Stadt-Welt. Über die Globalisierung städtischer Milieus (Die Zukunft des Städtischen). Frankfurt, New York: Campus

March, James G.; Olson, Johann P. (1990): Die Unsicherheit in der Vergangenheit: Organisationales Lernen unter Ungewissheit. In: March, James G.; Olson, Johann P. (Eds.): Entscheidung und Organisation. Kritische und konstruktive Beiträge, Entwicklungen und Perspektiven. Wiesbaden

Mayntz, Renate (1985): Soziologie der öffentlichen Verwaltung. Heidelberg: C.F. Müller

Morgan, Gareth (1993): Imaginization – the art of creative management. Newbury Park, CA: Sage

Ogburn, William F. (1964): Cultural Lag as Theory. In: William F. Ogburn (Ed.): On Culture and Social Change. Chicago: University of Chicago Press , pp. 86-95

Orlikowski, Wanda J. (1992): The Duality of Technology: Rethinking the Concept of Technology in Organizations. In: Organization Science 3/92, pp. 398-427

Ortmann, Günther; Sydow, Jörg, Windeler, Arnold. (1997): Organisation als reflexive Strukturation. In: Ortmann, Günther; Sydow, Jörg; Türk, Klaus (Eds.): Theorien der Organisation. Die Rückkehr der Gesellschaft. Opladen: Westdeutscher Verlag, p. 315-355

Rogers, Everett (1983): The Diffusion of Innovation. New York: Macmillan

Sassen, Saskia (1994): Cities in the World Economy. Thousand Oaks, California: Pine Forge

Weingart, Peter (Ed.) (1989): Technik als sozialer Prozess. Frankfurt/M.: Suhrkamp

Wilkesmann, Uwe (1999): Lernen in Organisationen. Die Inszenierung von kollektiven Lernprozessen. Frankfurt a.M./New York: Campus

Woolgar, Steve (2002): Virtual Society. Technology, Cyberbole, Reality. Oxford: Oxford University Press

8 Measuring Beliefs About ICT: The Case of European Urban Front Liners

Galit Cohen-Blankshtain and Peter Nijkamp

1. Prologue to urban dynamics

We are moving towards an urban world. Our geographical landscape is characterized by interconnected urban areas displaying complex spatial network structures. The recent emergence of information and communication technologies (ICT) has relaxed the bottlenecks imposed by the 'tyranny of distance', but has not eliminated spatial frictions. The popular idea of the 'death of distance' is certainly not supported by an overwhelming quantity of empirical evidence. To the contrary, in many countries we even witness a reinforcement of urban functions. Apparently, the agglomeration externalities are stronger than the potential for spatial dispersion that is made possible by the ICT sector. Clearly, we observe ICT-instigated phenomena such as telecommuting, tele-shopping, tele-education and even e-cities, but the final result of all such forces is not an erosion of the urban economy, but rather a more pronounced position in a global network setting.

As argued elsewhere (see Cohen et al., 2002), many modern urban policies are driven by myths and beliefs, and not by facts. Nevertheless, non-verified and sometimes speculative policy perspectives of the new economy decisively determine the future pathways of our cities. To identify the beliefs of urban front liners and to map out the relationship between their beliefs and policy decisions requires a thorough attitudinal and perceptional analysis based on experimental fieldwork regarding ICT.

ICT are expected to have various socially beneficial qualities. They are expected to mitigate many social problems: e.g. economic growth, education, efficient governing, public participation, social gaps and traffic congestion. Since ICT initiatives are relatively new, evidence regarding the success of ICT in those different areas is still limited and consequently the knowledge about the actual effects of ICT is still immature (Cohen et al, 2002).

Many scholars claim that following deregulation of ICT at national level, local government has a major role in the ICT area (Gibbes and Keite, 1997; Graham, 1997; Graham and Dominy, 1991). However, as Graham (2002), Graham and Marvin (1996) and Gibbs and Tanner (1996) have demonstrated, ICT policy has been considered mainly at a national rather than at a local, urban level. Nevertheless, in many cities local authorities have recognized the importance of ICT (Gibbes and Tanner, 1995) and the information revolution.

Adoption of urban ICT initiatives depends on the perceived benefits and costs that are associated with such initiatives. One factor that may affect these perceived costs and benefits is the attitude and belief of urban policymakers about ICT. Therefore, this study concentrates on urban front liners and their beliefs about ICT. We aimed at developing a framework of ICT beliefs and identified major constructs and measures associated with this framework. In Section 2 we present the background for this study and the importance of evaluating attitudes and beliefs about new technologies. Section 3 reviews literature on evaluations of personal convictions about computers and Internet as the departure point for developing a model to evaluate the convictions of urban policymakers about ICT in general. Next, Section 4 presents the conceptual model, and Section 5 describes the data used in the study. Section 6 shows the model estimated and model fit indexes and Section 7 discusses the results and concludes this study.

2. Study background

Technological developments are strongly associated with human progress. Already 2500 years ago, the Greek philosopher Xenophanes said: *"The gods did not reveal from the beginning all things to mortals, but through time, by searching, mortals discover what is best"* (Xenophanes, cited in Hadjilambrinos, 1998). Whether these developments are manna from heaven or the result of human creativity is still a disputed matter, but there is no doubt that technological developments had, and still have, a major impact on the way our society has evolved, and are likely to continue their critical role.

At face value, ICT is a collection of technologies and applications, which enables the processing, storing and transfer of information for a wide variety of users or clients. The collection of these technologies is expected to have a major impact on society. Therefore, several authors have labelled the current or forthcoming ICT-dependent developments as the "Information Revolution" or the "Information Age" (see, e.g. Castells, 1996; Slack and Fejes, 1987; Friendries, 1989). Negroponte (1995) even claims that we are already in the post-information age[1]. This view is the basis for many expectations of radical changes anticipated in urban areas.

Throughout history, there have been several approaches to assess the effects of technological changes on society and to predict how new technologies will affect society. In the context of ICT, Graham and Marvin (1996) identified two main approaches:

(1) Technological determinism tends to assume simple linear relations between technological causes and urban effects. This approach is based on a linear notion that innovation leads to new technologies that – when applied – have an impact on society. In other words, the hidden assumption in those predictions is that, if technology has the *ability* to affect the city in a certain way, it *will* affect it.

There are two main streams of technological determinism: Utopianism-futurism and Dystopian/urban political economy.

[1] He claims that in the Information Age there were still economies of scale, while in the post-Information Age the main feature is individualism.

Utopianism-futurism predicts the technological future of cities, households and society. It tends to take a relatively optimistic view of the future impact of telecommunications on cities and urban life. Where negative effects emerge, they can often be solved through new technologies.

Dystopian/urban political economy relates the application of ICT to society and the capitalist world. ICT supports a wider process of economic, political and spatial restructuring. The neo-Marxist stream also argues that ICT implementation is driven by the economic forces of capitalism and the need to overcome crises that reflect its inherent contradictions.

Both the utopian and the dystopian approaches attach normative aspects to the new technologies. The technologies are 'good' or 'bad' by their very nature. In general such deterministic views are associated with literature that predicted the effects of ICT before it was widely adopted. However, it seems that the attitude to those new technologies is changing due to the fact that, as the 'future' comes closer, it also becomes more realistic. Naturally, predictions about a completely new technology suffer from a higher degree of uncertainty (and tend to be more extreme and simplistic) than predictions about technology already in use. Unlike the 1980s and early 1990s (when most of the deterministic approaches took place), technologies are now in place or are rapidly adopted, but the spatial consequences are not immediately apparent. Thus, attitudes towards ICT and expectations about its influences are not totally divorced from the society that is adopting it. Therefore, nowadays a more complex approach to ICT is taken:

(2) Social and political construction of technology is a departure from these two deterministic approaches. It sees technology as part of society and determines how human, social and political processes can shape the way in which ICT is developed and applied in society. Individuals, social groups and institutions are seen to have a degree of freedom in shaping the development and applications of technologies in specific cases.

As Salomon (1998) stresses, technology, in general, and information technology in particular, is not merely a collection of hardware. Technology is a social construct. Moreover, information and communication technologies are neutral and may, depending on specific contextual factors, accommodate either dispersion or concentration trends (Nijkamp and Salomon, 1989). Moreover, the use of the technology is on one hand regulated by the value it provides to the potential user and, on the other hand, by various rules and norms, and these uses and regulations determine the impacts of ICT on society and its spatial structure. According to this approach, the new technologies are value free. The way they are adopted by individuals and society is largely determined by their value (Hadjilambrinos, 1998).

Gradually IT technologies and applications are entering many aspects of life, modifying and changing the way we work, manufacture, shop, study, make social contacts and other human activities. The more these technologies are adopted, the more we realize the differential effects of ICT on different aspects of life. While ICT is expected to have minor effects on one aspect, it is expected to have major effects on others. ICT applications can substitute certain activities, but it may also generate activities, modify others, or, in other cases it may have no affect at all (Mokhtarian and Meenakshisudaram, 1999). Therefore, alongside studies that aim at investigating the overall effects of ICT on society, more and more studies are focusing on specific aspects and studying the possible effects on these technologies on these particular aspects (i.e. substitution, generation, modification and neutrality). Studies that investigated the relationships between ICT and travel (Salomon, 1986; Mokhtarian, 1991; Nijkamp and Salomon, 1989) led to studies that focus on different travel purposes, and their associated tele-activities (e.g., Mokhtarian, 1998; Garrison, and Deakin, 1988 about tele-working; Nagurney et al, 2001 about tele-shopping, Lock et al, 2003 about tele-education). These studies indicate that each travel purpose can be affected differently by the same technology. As a consequence, the efforts to predict future impacts of ICT focus on its influence on various human activities, and not on the potential of the technologies per se. In other words, the same technologies can have different impacts on different activities and therefore it is important to study the expected influences on each activity. As the technologies are penetrated and used rapidly, our experience with the various applications shapes the way we evaluate the technologies and our expectations for further developments.

For more than twenty years, researchers and planners in many countries have been trying to hypothesize on, and to predict the effects of ICT on the city. Ideas about the role of the city, its features or even the question of whether it will actually survive have been published in many visionary books and papers. These complex relationships have led to many metaphors, which try to capture the futuristic and far-reaching consequences of ICT for the city and society at large. We can find metaphors like the knowledge-based city, the barrier free city, electronic space, the intelligent city, the electronic cottage, the flexicity, the telecity, the non-place urban realm, the idea-based city, the communications city, the virtual city, teletopia, the binary city, the invisible city, the wired city, and the network city (Graham and Marvin, 1996; Grosveld, 2002). The large number of metaphors indicates the expectation of important and significant effects of ICT on the city and its society.

Urban front liners may agree or disagree on these varied visions, but surely cannot ignore them when policies for the future have to be developed (Cohen et al, 2002). Moreover, the ICT discourse does not take place solely in the academic field. On the contrary, speculations about future urban influences of ICT also exist in the popular media, literature and art. Expectations and concerns about ICT are expressed in many respects (economic growth, social segregation, environmental issues and so on) and may motivate urban policymakers to employ different policies related to ICT. Thus, the intensive ICT debate raises the question on how urban policymakers assess the opportunities arising from ICT policies and the relevance of such policies for their city.

3. Beliefs about ICT

A large body of literature regarding the city (and society) and ICT is to be found. Most of this literature attempts to analyse ICT as a shaping factor or, in other words, it tries to explore ICT's possible spatial and social implications. However a more policy oriented motivation for such studies is to examine the ways in which intended ICT policies can (or would) be part of the ICT forces that will shape the future. Clearly, these two issues are related and affect each other.

On the one hand, when one is trying to evaluate, hypothesize or forecast future influences of ICT, one should also take into account possible implications of policies that are related to ICT. These policies (or the absence of policies) may affect the way ICT could shape the city and society at large. On the other hand, ICT policies aim to accommodate (present and) future needs and developments. Therefore, initiating relevant policies should be based on an evaluation of the possible trends and expectations of these technologies. In this study, we are interested in the way urban front liners perceive the future impacts of ICT. However, the speculations about the future and the role of the new technologies are value free. As Geels and Smith (2000) stress:

"Expectations may be biased by the broader cultural concerns of the time. People's perception of the future is often coloured by their cultural lens. Consequently, future expectations often reflect broad cultural concerns and hopes. This, to some extent, may also explain why certain future images gain widespread popularity" (page 877).

The speculations, forecasts and evaluations of the impacts of ICT on society, and on cities in particular, may affect the perceptions and beliefs of urban front liners. Their beliefs are likely to affect their evaluation of the relevance of ICT policies to their cities. In a similar way to the personal adoption of new technologies, decisions to adopt a *policy* regarding new technologies, are affected by beliefs about that technology.

Both the adoption and penetration of technologies are affected by the acceptance of the technology. Social psychology scholars developed the Technology Acceptance Model (TAM), which explained user acceptance of a technology based on user attitudes. TAM suggests two specific behavioural beliefs: perceived ease of use and perceived usefulness. These perceptions determine an individual's behavioural intentions to use technologies (Liaw, 2002). Based on this general technology acceptance model, specific models for computer and Internet acceptance were developed (Kay, 1989; Loyd and Loyd, 1985; Ledere et al, 2000). Following the extensive literature on perceptions of technologies in general, and computers and the Internet in particular, we offer a model to measure perception of ICT. Moreover, since we are interested in adoption of urban ICT policies, we are interested in the perceptions of urban decision makers (both elected politicians and administrative staff). In addition to the perception of personal usage and usefulness of ICT, we would like to pay attention to perception of the usefulness of ICT to the city and to its administration. Therefore, we use the TAM and other models that are based on theory of reasoned action (Fishbein and Ajzen, 1980) as a platform to measure and analyse perceptions of public decision makers. We suggest that these perceptions are likely to affect their willingness to promote urban ICT policies. We argue that adoption of technology is analogous to adoption of policies regarding new technologies. Beliefs about ICT affect perceptions of the relevance of ICT policies (Cohen-Blankshtain et al, forthcoming) and therefore affect the willingness to initiate and adopt ICT policies. For that reason this research aims at defining beliefs about ICT. Specifically, we are interested in conceptualizing and measuring beliefs about different aspects of ICT, and whether the views and beliefs of different aspects of ICT are correlated.

4. Modelling beliefs about ICT

Social scientists are often interested in exploring theoretical constructs that cannot be observed directly and can only be approximated by observable or measurable variables. These abstract phenomena are termed "latent variables" (Byrne, 1998). The creation of latent variables is based on a confirmatory factor analysis, where we can specify which variables define each factor (i.e. latent variable). The observed variables are as well termed "indicators, since they indicate the latent constructs (Hair et al., 1998). Beliefs about ICT can be regarded as a latent construct since it is a multidimensional concept and therefore cannot be observed directly. For that reason, we use various indicators that measure beliefs about ICT. Moreover, we claim that the multidimensional feature of beliefs about ICT requires a second order latent variable, i.e. a latent contrast that is built upon latent constructs from a lower level.

We consider three relevant aspects of the way ICT is perceived by urban policy makers. The first aspect concerns the perceived influences of ICT on the attractiveness and strength of their city. This aspect may be regarded as the usefulness of ICT from an urban perspective.

The second aspect involves the perceived influences of ICT on the urban administration and its relation with its citizens. This aspect can be seen as the usefulness of ICT from an administrative and policymaking perspective. The third and last aspect is the usefulness of ICT from an individual perspective. The three aspects are reconsidered to be first-order latent variables since each of the aspects is built on a number of indicators. The variable that represents the perceived influences of ICT on the city is based on four indicators. The variable that represents the perceived influences of ICT on the city administration is based on seven indicators and the third variable that represents the individual usefulness of ICT is based on four indicating variables. There are various indicating variables that may represent these three aspects. After a process of elimination of variables with low-item-construct scores, fifteen indicating variables were chosen. Figure 1 illustrates the final indicators that were chosen to represent each aspect. The variable that represents the general beliefs about ICT is based on these three aspects. Therefore, the general ICT beliefs are represented by a second-order latent variable that is constructed by these three first order latent variables and does not have direct indicating variables.

The hypothesized overall model for beliefs about ICT is presented in Figure 2. It represents a second-order latent variables model. The model hypothesizes a priori that: beliefs about ICT can be conceptualized in term of three factors; each observed variable will have a none zero loading on the factor it was designed to measure, and zero for all the other factors; and, covariance among the three first-order variables will be explained by their regression to the second-order factors. In the Figure, the single-headed arrows leading from the second-order factor of ICT beliefs (F4) to each of its underlying first-order factors (F1 to F3) are regression paths that indicate the predictions of beliefs about ICT and the city (F1), beliefs about ICT and city governing (F2), and satisfaction from ICT usage (F3) from a higher order ICT beliefs factor. They also represent second-order factor loads. Each first-order factor also has a residual disturbance term (DE1, D2, D3), which represents errors in the prediction of the first-order factors from the higher order factor. Finally, each first-order factor is predicted by its manifested (observed) variables (V1 through V15).

Figure 1: First-order latent variables and their indicators

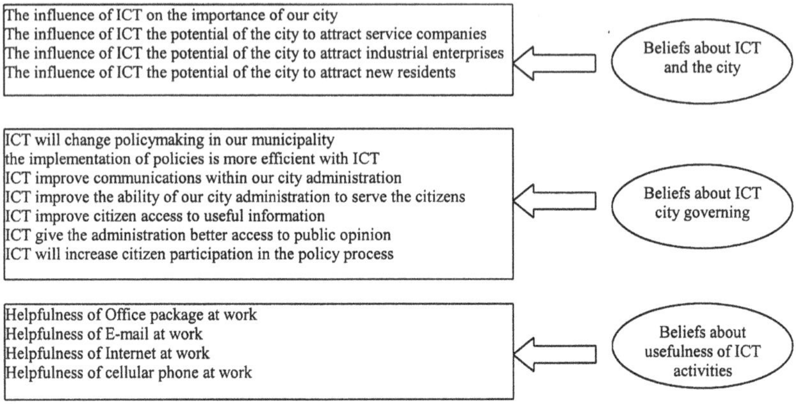

The observed variable, which indicates the latent constructs under investigation were chosen in an iterative way. Starting from a large number of indicators, indicators that had low scores with their latent construct were eliminated, ending with 15 variables that represent their constructs in the best way.

As the model suggests, there are three factors that operate as independent variables, each of which can be considered as one dimension of beliefs about ICT. However, we argue that there is a higher-level factor that is considered accountable for the lower-order factors. The higher-level factor is hypothesized as accounting for the variance and covariance related to the first order-factors. The higher-order factor does not have direct observed indicators, but it is linked indirectly to the variables measuring the lower order-factors.

Figure 2: Schematic classification of model of beliefs about ICT

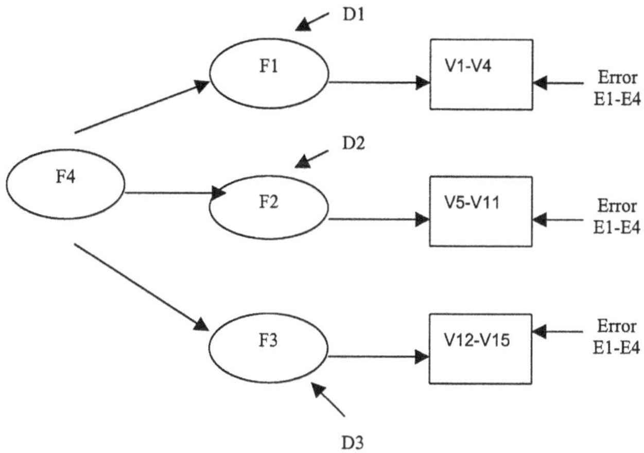

F1= first order factor for beliefs about ICT and the city
F2=first-order factor for beliefs about ICT and urban governing
F3=first-order factor for satisfaction from ICT usage
F4=second-order factor for overall beliefs about ICT
V1-V15=observed variables

5. The database

As part of the European project TeleCityVision[2], an extensive survey was conducted, targeting urban decision makers (both politicians and responsible administrative staff) in more than 200 cities in seven European countries (Austria, France, Germany Ireland, Norway, the Netherlands, and Spain). The survey was conducted between May and September 1999. The questionnaires were sent to various municipal departments that were supposed to have a direct or indirect influence on ICT-related activities in the city, as well as to elected officials of the city (politicians). The effort to include various municipality department members in our sample was due to the fact that ICT policies

[2] TeleCityVision-Project was funded by the European Commission and co-ordinated by BIS; participants are: BIS (Germany), COMTEC (Ireland), ESI (The Netherlands), ETSI-UAM (Spain), ICCR (Austria), THEMA (France), and ZTG (Germany)

and strategies are not made by one single recognized responsible body.

As one might expect, the willingness to answer these questionnaires varied across the countries; there is an overrepresentation of German and Norwegian respondents (who were most responsive to the questionnaire). However, since this study focuses on the individual level and not on comparing cities, this bias was not considered problematic. Approximately 1500 responses were returned (the response rate varied between 20 and 40% across the countries), half of them from Germany. The respondents were from 239 cities in Europe. The size of cities that are represented in our database is shown in Table 1.

Table 5.4: Respondents according to their city size

Size of the city	Percentage from the sample
up to 50,000	15.8
50,001-100,000	11.1
100,001-200,000	32.4
200,001-500,000	22.3
500,001-700,000	8.5
700,001-1,000,000	3.8
1,000,001-1,500,000	.8
1,500,001-2,000,000	2.1
2,000,001-3,000,000	.6
More than 3,000,000	2.0
Missing	.6
Total	100.0
Total	100.0

Among the various urban front liners that were sampled in the survey, the largest group in the sample is the urban planners (32.5%). The second largest group consists of economists (15.8%). The third largest group is the elected politicians (15.3%), who speak for the public and are supposed to approve policies that are implemented by the administrative bureaucracy. The rest of the respondents are active in various administrative functions. The education level of the respondents is relatively high. The vast majority of the respondents in our sample have at least a Bachelor's university degree and most of the respondents are in the 40-60 age group.

As Table 2 shows, most of the respondents use various ICT applications, and therefore have some ICT experience.

Table 2: Usage of different ICT applications

Application	Daily	Weekly	Less than weekly	Not available	No answer
PC	84.3%	4.6%	2.1%	5.2%	3.3%
Word processing	80.9%	8.3%	4.4%	2.7%	3.7%
Databases	33.9%	19.9%	28.8%	9.1%	8.2%
E-mail	61.7%	11.7%	9.1%	12.7%	4.7%
Internet	40.3%	19.1%	12.6%	22.2%	5.9%
GIS	9.5%	9.1%	20.9%	46.9%	13.5%
CAD	8.1%	3.6%	17.1%	57.7%	13.3%

6. Testing the hypothesized model

In confirmatory factor analysis, the purpose is to test whether the hypothesized construct is indeed confirmed by the data. Our model was estimated with a Lisrel model that enables testing confirmatory factor analysis and creating of latent variables (Joreskog and Sorbom, 1996). There are several statistical tests that can indicate whether the observed indicators do, indeed, correctly represent the hypothesized latent construct. The first step is checking for 'uni-dimensionality', i.e. the extent to which each indicator has an acceptable fit with its factor and whether the estimated factor loadings are statistically significant. In our model all the indicators are significant at the .01 level. Other statistical tests aim at measuring the reliability and validity of the indicators as representing a common factor.

Figure 3: Estimation of the model of beliefs about ICT

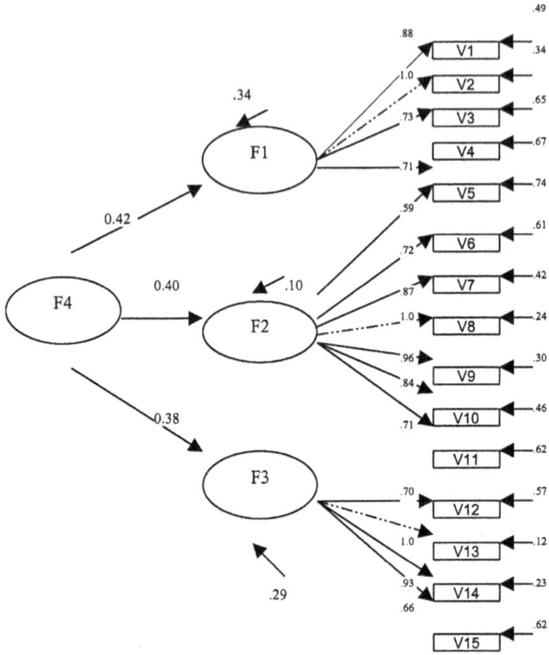

V1=observed variable: the influence of ICT on the importance of our city
V2=observed variable: the influence of ICT the potential of the city to attract service companies
V3=observed variable: the influence of ICT the potential of the city to attract industrial enterprises
V4=observed variable: the influence of ICT the potential of the city to attract new residents
V5=observed variable: ICT will change policymaking in our municipality
V6=observed variable: the implementation of policies is more efficient with ICT
V7=observed variable: ICT improve communications within our city administration
V8=observed variable: ICT improve the ability of our city administration to serve the citizens
V9=observed variable: ICT improve citizen access to useful information
V10=observed variable: ICT give the administration better access to public opinion
V11=observed variable: ICT will increase citizen participation in the policy process
V12=observed variable: helpfulness of Office package at work
V13=observed variable: helpfulness of E-mail at work
V14=observed variable: helpfulness of Internet at work
V15=observed variable: helpfulness of cellular phone at work
—··—··— Represents variable that is used as scaling variable, for identification purposes

Cronbach's alpha measures the reliability of construct indicators. Its values range between 0 and 1.0, with higher values indicating higher reliability among the indicators (Hair et al, 1998). It is recommended to have a value of at least .7. Cronbach's alpha for F2 and F3 is higher than .7 (.82 and .72, respectively) while for F1 it is equals .69, slightly less than the recommended value.

A related test is the construct reliability, which is the degree to which a set of indicators is consistent in measurement. The measure is calculated as:

$$\frac{\sum (\lambda^2)}{\sum (\lambda^2) + \sum (\varepsilon)}$$

where λ is the vector of the standardized loadings, and ε is the indicator measurement error. The construct reliability should exceed .50 to indicate a good representation of the construct (Hair et al, 1998). Here again, the reliability of F2 and F3 is satisfying (.52 and .62, respectively) while for F1 it is lower than the recommended value (.46).

The final test is the variance-extracted measure. This measure reflects the overall amount of variance in the indicators accounted for by the latent construct.

Higher values represent a greater degree of shared representation of the indicators by the latent variable. The index is calculated as:

$$\frac{(\sum \lambda)^2}{(\sum \lambda)^2 + \sum (\varepsilon)}$$

Here also, λ is the vector of the standardized loadings, and ε is the indicator measurement error. Guidelines suggest that the variance-extracted value should exceed .50 for a construct (Hair et al, 1998). In our case, the variance-extracted value for all factors is high and therefore suggests good representation. Although factor F1 scored slightly low values in the first two tests, its score in this test is satisfactory, and therefore we conclude that F1 is represented at an acceptable level by its indicator.

Table 3: Fit of latent constructs F1, F2 and F3

ICT and the city (F1)			ICT and governing (F2)			ICT satisfaction (F3)		
Indica-tor	Factor load [a]	t-value	Indi-cator	Factor load [a]	t-value	Indi-cator	Factor load [a]	t-value
V1	.72	Na	V5	.51	Na	V12	.66	Na
V2	.81	24.3	V6	.63	37.2	V13	.94	9.9
V3	.59	20.6	V7	.76	35.4	V14	.89	9.6
V4	.58	21.2	V8	.87	37.8	V15	.62	6.1
			V9	.83	36.9			
			V10	.73	37.1			
			V11	.61	35.0			
Cronbach's alpha = .69 construct reliability = .46 variance extracted = .77			Cronbach's alpha = .82 construct reliability = .52 variance extracted = .88			Cronbach's alpha = .75 construct reliability = .62 variance extracted = .86		

a - standardized factor loads

** Significant at the 99% level.

Various goodness-of-fit measures have been developed to evaluate the correspondence of the actual (observed) input matrix with that predicted from the proposed model. Since model fit is a multifaceted concept, each measure indicates a different aspect of the model and the extent to which, according to this measure, the model fits well.

As the various fit indexes measure different aspects of the model, they may inspect problems in some aspects, and indicate a good fit in others. As Table 4 shows, most of the fit indexes are either good or slightly lower than the recommended level. Therefore we may conclude that the model represents the data in a satisfactory way.

Table 4: Fit indexes

Measure	Short explanation	Recommended value	Model value	Comments
Chi-square	A poor fit measure. Tests the extent to which the model specification is true.	Should be significant; $\chi^2/df < 3$.	$\chi^2=485.59$ df=87 $\chi^2/df=5.5$	Unrealistic in most SEM empirical research
RMSEA Root Mean Square Errors of Approximation	Error of approximation in the population.	Good fit: <.05 Reasonable fit: <.08	.057	
NFI Normed Fit Index	Derived from the comparison of a hypothesized model with an independence model, and indicates the proportion in the improvement of the overall fit of the model relative to a null model.	> .90	.89	Affected by sample size
CFI Comparative Fit Index	Derived from the comparison of a hypothesized model with an independence model, and indicates the proportion in the improvement of the overall fit of the model relative to a null model.	> .90	.89	Corrects the effects of sample size from NFI
RMR Root Mean Square Residual	Average covariance residuals: the differences between the observed model and model-implied covariance.	<.10	.066	
GFI Goodness-of-fit Index	Proportion of the observed covariance explained by the model-implied covariance.	>.90	.89	
PGFI Parsimony Goodness-of-fit Index	Proportion of the observed covariance explained by the model-implied covariance that takes the complexity of the hypothesized model into account.	>.50	.63	

7. Discussion and interpretation of results

The first set of relationships examines the fit between the measures and the constructs they were intended to capture. As the three construct measures in the previous section

indicate, we may indeed conclude that the three aspects of beliefs about and perceptions of ICT are represented well by its indicators.

Next, we can see that the relation between all first-order and second latent variables is significant. Moreover, the variance extracted from F4, the second-order latent variable, equals 0.84. It implies that much of the variance in F1, F2 and F3 can be captured by F4. The second-order model indicates that indeed the three different aspects of ICT perceptions are relevant components of the overall ICT concept. It also points out that the different aspects of ICT beliefs should all be included in order to represent a comprehensive evaluation of ICT. Concentrating on just one aspect is, according to the model, insufficient to evaluate the complex variable referred to as ICT perceptions.

The model confirms our argument that perceptions and attitudes towards ICT have various aspects but are all related to fundamental concepts about these technologies. Therefore, although we have identified three distinct belief aspects about ICT, they are still determined (to a certain degree) by a more fundamental and general attitude towards it.

The advantage of using the second-ordered variable that represents urban front liners' beliefs about ICT is that it can be used as a single variable in any statistical analysis that seeks a relationship between belief in ICT and the initiation of ICT policies. In order to test the applicability of the theory of reasoned action (Ajzen and Fishbein, 1980), there is a need for an operational variable that captures beliefs and attitudes. This paper contributes to a framework, which helps to measure the somewhat vague concept about ICT. Based on models for personal adoptions of new technologies, we offered an extended concept that links beliefs about technologies to the willingness to adopt policies, which use these technologies. Nevertheless this is a preliminary step towards testing the relationships between ICT beliefs and the willingness to initiate ICT policies. Future research about these relationships may use the framework in this study to conduct the study.

References

Byrne, B.M., 1998, Structural Equation Modeling with LISREL, PERLIS and SIMPLIS: Basic Concepts, Applications and Programming, Lawrence Erlbaum Associate, Publishers, Mahwah NJ.

Castells, M., 1996, The Rise of the Network Society, Blackwell, London.

Cohen, G., Salomon, I. and Nijkamp, P. 2002, Information-Communications Technologies (ICT) and Transport Policy: Does Knowledge Underpin Policy?", Telecommunication Policy, 26(1-2), 31-52

Fishbein, M. and Ajzen, I., 1980, Understanding Attitudes and Predicting Social Behavior, Prentice Hall, Englewood Cliffs, NJ.

Friendreis, J., 1989, "The information revolution and urban life", Journal of Urban Affairs, 11(4), 327-337.

Garrison, W. and Deakin, E., 1988, "Travel, work, and telecommunications: a long view of the electronic revolution and its potential impacts", Transportation Research, 22A(4), 2390-2246.

Gaspar, J. and Glaeser, E.L., 1998, "Information technology and the future of cities", Journal of Urban Economics, 43, 136-156.

Geels, F.W. and Smit, W.A., 2000, "Failed technology futures: pitfalls and lessons from a historical survey", Futures, 32, 867-885.

Geenhuizen, M. van and Nijkamp, P., 2000, "Electronic banking and the city-systems in the Netherlands, in Leinbach, T. and Brunn, S.D (eds.), Worlds of E-commerce: Economic, Geographic and Social Dimensions, John Wiley, New York, 181-201.

Gibbs, D. and Tanner, K., 1995, "Communications technology, local economies and regulation theory", TESG, 88(1), 29-40.

Gibbs, D. and Tanner, K., 1996, "Information and communication technologies and local economic developments policies: The British case", Regional Studies, 31(8), 765-774.

Graham, S. and Marvin S., 1996, Telecommunications and the City. Electronic Spaces, Urban Places, Routledge.

Graham, S., and Dominy, G.R., 1991 "Planning for the information city: The UK case", Progress in Planning, 35(3).

Graham, S., 1997, Telecommunications and the Future of Cities: Debunking the myths", Cities, 14(1), 21-29.

Grosveld, H. 2002, The Leading Cities of the World and their Competitive Advantage. The Perception of 'Citymakers', The Netherlands, World Cities Research.

Hadjilambrinos, C., 1998, "Technological regimes: An analytical framework for the evaluation of technological systems", Technology in Society, 20, 179-194.

Hair, J.F., Anderson, R.E., Tatham, R.L. and Black, W.C., 1998, Multivariate Data Analysis, Fifth Edition, Prentice Hall, N.J.

Joreskog, K.G. and Sorbom, D., 1996, Lisrel 8: User Reference Guide, SSI International.

Kay R.H., 1989, "A practical and theoretical approach to assessing computer attitude measure (CAM)", Journal of Research on Computing in Education, Summer, 457-463.

Liaw S., 2000, "An Internet survey for perceptions of computers and the World Wide Web: relationships, predictions, and difference", Computers in Human Behavior, 18, 17-35.

Lederer A. L., Maupin D.J., Sena M.P. and Zhuang Y., 2000, "The technology acceptance model and the World Wide Web", Decision Support Systems, 29, 269-282.

Loyd B.H. and Loyd D.E., 1985, The reliability and validity of instruments for the assessment of computer attitudes", Educational and Psychological Measurement, 45, 903-908.

Mokhtarian, P.L. and Meenakshisudaram, R., 1999, "Beyond telesubstitution: disaggregate longitudinal structural equation modelling of communication impacts", Transportation Research Part C, 7, 33-52.

Mokhtarian P., 1991, "Telecommuting and travel: State of the practice, state of the art", Transportation, 18(4), 319-342.

Mokhtrian P.L., 1998, "A synthetic approach to estimate the impacts of telecommuting on travel", Urban Studies, 35(2), 215-41.

Negroponte, N., 1995, Being Digital, Hodder and Stoughton, London.

Nijkamp, P. and Salomon, I., 1989, "Future spatial impacts of telecommunications", Transportation Planning and Technology, 13, 275-287.

Salomon, I., 1998, "Technological Change and Social Forecasting: The Case of Telecommuting as a Travel Substitute", Transportation Research Part C, 6, 17-45.

Salomon I., 1986, "Telecommunications and travel relationships: a review", Transportation Research A, 20A(3), 223-238.

Slack, J.D. and Fejes, F. (eds.), 1987, The Ideology of the Information Age, Norwood, Ibex, New Jersey.

7 Network City and City Networking
A Local Administration Actors Perspective in Seven Western European Countries

Manuel Valenzuela, Carmen Vázquez, Santiago Lorente

Abstract

After dealing with the theoretical background of the concepts of network city and city networking the paper descrives the objectives, hypotheses and methodology of the research. It tries to analyse the first results of a quantitative survey using the theoretical framework which connects urban form/functions and ICTs through a process of domestication of technology by the end users. In order to break down all the data collected a set of urban dimensions likely to be affected by the introduction of ICTs has been distinguished: city self image/urban marketing, administrative dimension, participatory dimension, infrastructural dimension, planning dimension, city networking and economic dimension. And last but not least, the fourth part will concentrate on the process and policies dealing with city networking, a controversial point having in mind the ambiguous perceptions about the relations between urban competition and cooperation.

1. The concepts of network city and city networking: a theoretical approach

The concepts of network city and city networking have been a fruitful arena of debates for urban scholars along the last decades. The former focuses on the new pattern of urban infrastructures and relations which is changing the form and functioning of our cities as dramatically as railroads, highways, electric power supply, and telephone networks did in the past. It has been stated with a certain amount of reason that "Intraurban digital networking furthers the long evolution of human settlements from loose collections of more or less independent dwellings to highly integrated, networked cities in which multiple infrastructures of tracks, pipes, and wires deliver centrally supplied services to buildings and carry away waste" (Mitchell, 1999: 22). Nevertheless, as another urban scholars have pointed, until now we lack a critical networked urbanism due to "disciplinary failings and the neglect of networked infrastructures, their hidden and taken-for-granted nature, assumptions of technological determinism, and the panic effects of networked collapses" (Graham & Marvin, 2001: 30).

On the other side, but complementing the previous urban restructuring at a different scale, urban networking has became one of the modern "mottos" in urban studies. And, even though ICTs have increased the amount of contributions on this subject, some scholars highlight that "the external networks are not all infrastructure. They consist rather of relationships of a less tangible nature with many outside places, mostly other cities. It is however often forgotten that in nature and essence a city is an organization of human beings acting together, hence an artifact society. Networks between cities became the essence of history" (Gottmann, 1991: 277).

"Cities exist in a world of flows, linkages, connections, and relations. World cities represent an alternative metageography, one of networks rather than the mosaic of states. Historically, cities have always existed in environments of linkages, both material flows and information transfers. They have acted as centers from where their hinterlands are serviced and connected to wider realms" (Beaverstock et al., 2000: 123). Following the previous statements it is possible to guess two of the main generic features across all the different research schools within the literature: the concentration on the attributes of "world/global cities" and the neglecting of their relations, in conclusion, studies are generally full of information that facilitates evaluations of individual cities and comparative analysis of several cities, but they lack on information about the links. Recently, empirical studies that present networks of world cities have focused on international airline-passenger statistics (Keeling 1995; Kunzmann 1998;

Rimmer 1998), global office location strategies (Beaverstock et al. 2000) or flow data provided by Federal Express Corporation that measure the participation of different metropolitan centers in international information flows (Mitchelson and Wheeler, 1994). Even more recent research has shown how existing urban hierarchy is playing an important role in Internet content production, but even if the actual production of Internet contents depends of certain cities, it is also connected to a global informational network (Zook, 2001: 1694).

In the European context, the significance of spatial structures is undergoing fundamental changes. Despite the fact that it is assumed that cities in the "less-favored regions" enjoy the same footing as existing centers in terms of their potential access to the information and communication networks, other authors have warned about the rise of a "wealthy archipelago of city regions ... surrounded by an impoverished *lumpenplanet*" (Petrella, 1995: 21). In this sense, Petrella sets out his global apartheid dystopia as a warning about current trends so as to alert us to the dangers ahead. Anyway, we must bear in mind that world cities are not eliminating the power of states, they are part of a global restructuring which is "rescaling" power relations, in which states will change and adapt as they have done many times in previous restructurings. Along with the role of states we have assumed that the administrative willingness to innovate will largely decide how the future might be designed across different regional contexts in Europe. In this process, the communal actors of urban development, such as urban planners, economic promoters, and politicians concerned with urban development will play an important role. "How they perceive the processes by which information and communication media take hold, their own self-image, how they assess their latitude for action, and their strategies are all significant for the future shape of urban tasks" (TeleCityVision, Annex 1).

Municipal governments and urban development agencies are currently engaged in an increasingly pervasive competitive struggle to attract investment and to secure nodal positions for their cities on corporate global networks (Amin and Thrift, 1992; Harvey, 1989; Leitner, 1990; Swyngedouw, 1989). This is being driven by the increasingly powerful position which transnational corporations (TNCs) hold in shaping the fortunes of local economies. In the wake of urban economic restructuring and globalisation, there is now a remarkable homogeneity in the shift from what Harvey (1989) calls "managerial" service-oriented forms of urban governance to new strands of "urban entrepreneurialism". The construction of favourable urban images, or the reconstruction of unfavourable ones, is a central component of these new policies (Paddison, 1993). For investors, changes in communication technologies are increasing the economic uncertainty faced by cities; therefore, firms are becoming less dependant on particular cities and consequently they reconsider the attractiveness of their investment in a city and the possibility of relocation. This point of view must be seen together with the case of communication between plants and the opportunity of flexible production and technological change (Leitner and Sheppard, 1998: 291).

Academic and policy attention is beginning to switch to the possibility of supporting interurban collaboration as well as simple competition (Lovering, 1995). This is part of a wider exploration of new models of progressive "entrepreneurialism" in which the economic promotion of cities is matched with concern for social and spatial inequalities and the need for collaboration between cities (Leitner and Gardner, 1993). As well as emerging in a bottom-up fashion, as city authorities begin to see more scope for collaboration, certain national states and the widening range of urban projects at the European level are providing much added momentum to the exploration of interurban networks.

2. Setting the frame: description and methodology of the survey

Within the previous theoretical framework our contribution is part of a wider research focusing on uncovering and explaining both the potential for, and barriers to, innovation and development. We have assumed that "*the opportunities to innovate essentially depend on the strategies of actors in public institutions. The study examines how the actors of urban development assess the specific opportunities and risks in this process, and which strategies they use to meet them. How do they perceive the impact of the increasing use of information and communication technologies on the process of urban development?*

How do they integrate their vision for the future into concrete urban planning strategies?"[1].

The main goal of this contribution is to present a first approach to the findings of the quantitative inquiry, summarized in a general database which gathers the answers of urban actors of all major cities and selected Functional Urban Regions of the partner countries. During 2000 the seven research teams carried out a great quantitative survey through the mailing of 5,080 questionnaires to 275 cities all across Europe, collecting back 1,558 fulfilled questionnaires from 244 cities (response rate of 30%). One of the points worth to be noted is that within the frame of the database the German questionnaires are over-represented (more than 51% of the whole sample) which explains that our analysis is highly influenced by the German actor's vision. On the other side, the national/cultural bias can be overcome and complemented by using the department/professional attachment of the respondents and/or the size of the cities surveyed. Finally, for this contribution we have selected the city size (cities over 100,000 inhabitants – cities between 50,000 and 99,999 inhabitants – cities with less than 50,000 inhabitants) as independent variable because we really do not see significant national differences, on the contrary there are many similarities, and here we can distinguish between countries with lots of cities over a hundred thousand and those with more smaller cities. The creation of a more complex city typology according to the actors assessment, or in another words, a subjective city typology, has been the next step trying to go further on in deepening the analysis.

This paper reports preliminary results on the empirical groundwork required for analysing, interpreting and evaluating the complex relations between the perceptions of needs/possibilities expressed by a set of urban actors across seven European countries and the plans, practices and strategies they develop in order to meet the urban problems/potential of their own cities. The scale of analysis selected for

[1] The data used in this contribution are part of the results of the TCV Survey (2000). TeleCityVision working group (April 2000) "Questionnaire Survey" (unpublished) TCV project financed by the EC, coordinated by BIS (Germany), participants: BIS (Germany), COMTEC (Ireland), STS (Norway), ESI (The Netherlands), ETSIT-UAM (Spain), ICCR (Austria), THEMA (France) and ZTG (Germany).

this contribution is the individual one, because it cannot be forgotten that are working with visions and valuations of private people even though different municipal departments and areas fall under their scope. Anyhow, in order to facilitate the interpretation of data we have aggregated them in three city sizes, those initially used within the methodology of the survey, and one of the few objective variables of the data base. From the 244 cities surveyed, 182 have more than 100,000 inhabitants (74.6%), 47 between 50,000 and 99,999 (19.3%), and 41 less than 50,000 (16.8%). Moreover, we have tried to analyse the results at country scale, and lastly at city type scale, maybe one of the most problematic issues of the survey.

3. Basic aspects of network city and city networking

a. Meanings: perceptions of needs/possibilities

The answers classifying the cities in terms of rank/function have been a source of surprises for us. Almost all the cities display an over-estimated self image which reaches the peak in the case of big cities (over 100,000 inhabitants), in this group more than 35% define their city as a clear ICT and/or multimedia centre, while for the little cities the percentage decreases to 17.4%. The level of ICT usage is inconsistent; peaks of excellence are accompanied by troughs of incompetence. However, the "vision" of ours actors is pretty different: 47.3% of the Spanish questionnaires define their city as a clear ICT and/or multimedia centre, followed by the Austrian interviewees and the Irish ones, the most modest being Norway.

Bearing in mind the perception of the image and importance of their city displayed by the local actors interviewed it goes without saying that their considerations of the expected developments are equally optimistic. Almost all of them forecast an increase in importance for their cities, mainly linked to the potential for attracting service companies and processes of co-operation and competition between their city and other cities. The processes of urban competition are mainly assessed by local actors living both in big cities and in little ones and when we compare the answers to this group of questions with those to the next one: *...and how do you assess the influence of ICTs on these developments?* We will be able to guess to which extent they over- or underestimate the impact of the information society on the process of urban development. In general terms again the local actors living both in big/little cities attach a great importance to ICTs for those aspects linked to "urban importance", mainly for big cities, competition with other cities and the potential to attract service companies, or in other words, the processes related with tertiary activities and urban marketing. On the other hand, medium cities show a less enthusiastic approach to the idea that ICTs will increase the competitive position of the city. A learning deficit that supports the thesis of a knowledge gap among the administrative actors? Why specifically in medium cities and not so much in little ones? The answer is not clear at this point, but maybe it is worthy to note that among little cities the presence of Norwegian and Irish ones (both summarize 63.4% of the category) is remarkable. In the first case, we must remember that Norway is one of the most developed countries in the world in terms of telecommunication infrastructure, which is rather evenly distributed around the country.

Anyway, a more detailed break down at country scale yielded the following results: When they are asked *which of the following developments do you expect in the future?* Almost all of them forecast an increase in importance for their cities, mainly linked to the potential for attracting service companies and processes of cooperation and competition between their city and other cities. The processes of urban competition are mainly assessed by local actors in Germany and Norway while the Netherlands seem to face the future in a less competitive basis. This pattern is confirmed by the answers to the topic of future trends of co-operation between their cities and other cities. Now the most optimistic are France and the Netherlands while Germany displays again the most somber vision. If we compare the answers to this group of questions with those to the next one: *...and how do you assess the influence of ICTs on these developments?* We will be able to guess to which extent they over- or underestimate the impact of the information society on the process of urban development. In general terms the local actors attach a great importance to ICTs for those aspects linked to "urban importance", mainly for big cities, competition with other cities and the potential to attract service companies, or in other words the process related with tertiary activities and urban marketing. In this sense Ireland emphatically highlights the role of ICTs in urban competition while Germany keeps a more skeptical vision.

The answers to the topic of future trends of cooperation between their cities and other cities disclose the highest agreement level among medium cities. The local actors working in these municipalities show the most positive approach to the possibilities for creating, improving and reinforcing the ties among cities. In brief they seem to be the best candidates for evaluating the potential of ICTs in the city networking activities. However, bearing in mind this point the question: *as far as your municipality is concerned, to what extent do you agree to the idea that ICTs will support urban networking with another European cities?* try to give another screw to the perception of the interrelation between ICT and urban development, focusing on the estimated influence of urban actors on the diffusion of ICTs. The results emphasize firstly their connected valuation of urban networking and urban competition. The idea that ICTs will support urban networking with another European cities is especially defended by local actors from big cities while little cities keep a much more cautious profile. It looks like the position of the city in terms of rank-size and economic dynamism is the most valuable clue to forecast urban networking activities. The idea that ICTs will support urban networking with another European cities is specially defended by Austria and Ireland while only Norway risks to show its cautious disagreement. The potential of ITCs to increase the competitive position of cities within the European urban network is highlighted by Austria and Spain while the Netherlands keeps a much more conservative approach to this dimension of ICTs. Admittedly speculative, do they prefer trusting traditional physical infrastructures (Eurogates) and cultural/economic environment? It is a problem of learning deficit?.

As far as the infrastructural dimension is concerned, the question: *do you find the following applications helpful in your work?* directly tests the attitude of our local actors towards a handful of ICT applications at work. Assuming that all them show a meaningful appreciation of office package at work as well as of email the only noticeable nuances worth to be pointed out are that medium cities score over the average for Internet and email while local actors from little towns surprisingly display a strong dependency from cellular phones. Once more Germany performs the critical approach to the possibilities of the abovementioned applications and tools.

b. Doing: Plans and Policies

Switching from the meaning aspects (perceptions of needs/possibilities) and the participatory, administrative and economical dimensions to the plans and practices the different European cities developed, we observe a big gap among them. When we asked: *With regard to the ICT policies, which of the following statements is true for your municipality? My city is implementing a systematic and comprehensive ICT policy* the agreement is clear but not enthusiastic in the case of little towns, meanwhile big cities display the most critical vision, a question worth to be noted and able to be connected with the idea that private actors are often understood as the focal points in deploying urban information societies, less often the municipal or other authorities. The agreement is clear but not enthusiastic in the case of Austria, Norway and France, meanwhile The Netherlands display the most critical vision a question worth to be noted in a country with the highest level of Internet access. The collected answers to the statement: *My city is implementing an ICT policy which concentrates on specific areas/problems/needs*, attach the highest agreement level among medium cities, and again big cities cast some shadows on their supposedly innovating local urban policies. The highest agreement level goes to France, on the other hand Germany and again the Netherlands performs a more reluctant role.

The negative approach to the same topic works as a control variable trying to verify the consistence of the previous answers, or in others words, the response to the question: *My city has no specific ICT policies*, confirms the highest agreement among interviewees from big cities while actors from medium cities reject this statement. The highest agreement is confirmed among interviewees from Germany and the Netherlands while France clearly rejects this statement. And the last control question: *My city has no ICT policies at all*, discloses the polarized answer of actors from little towns which show a schizophrenic personality likely to be linked to awareness and knowledge gaps as well as to the inconsistence of ICT strategies; peaks of excellence are accompanied by troughs of incompetence. Big cities, traditionally the most critical about their ICT strategies, however openly rejected this statement. Ireland and Spain slightly agree and conversely they are openly rejected by France.

Figure 1. Implementation of systematic and comprehensive ICT policies by city size (%)

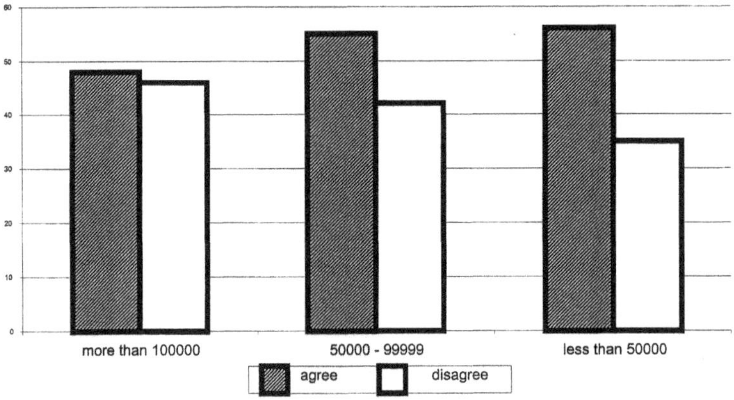

The sphere of action and politics is directly assessed for every city trough the question: *Is there a formal strategy plan/program on ICTs in your city?* Here the collected answers confirm the relative involvement of local governments in the case of big cities and the negative answer of interviewees living in little towns, but without the slightest doubt the most relevant feature in this question is the huge information gap in all the cases, if we consider the "don't know" responses are an indicator of low level of awareness, we must agree that a limited number of city actors knows of the various elements of ICT strategies and their implementation, moreover in none of the cities did all of the respondents know about these plans. Here the collected answers confirm the strong involvement of local governments in the case of France the low one in the Irish municipalities and surprisingly the information gaps in the case of Germany.

A more detailed portray of the Urban European policies dealing with the planning dimension, the administrative dimension, the participatory dimension, and the infrastructural dimension can be analyzed trough the assumed municipal goals. Except for the participatory dimension where big cities score the lowest positions within the sample, all the remaining goals selected reveal the hegemony of big cities: *improving telecommunication infrastructure* is at the top of the agenda in big cities; the second goal: *supplying municipality information via telecommunications networks (internet, teletex, telephone and so on)* seems to be in the priority list of big cities; *promoting municipality services via telecommunication networks* appears to be a key aspect for big cities. Finally, using ICTs in transport planning (intelligent road signs, online traffic information, etc.) fits mainly the politics of big cities and keeps a poor profile within those of little towns. Anyway, if these answers are compared with the city problems identified in the same questionnaire and we check the low consideration the traffic congestion received as far as the provision of ICT measures, it may be wondered if transport planning keeps an isolated management split up from that of physical planning. The first goal selected: *improving telecommunication infrastructure* is at the top of the French agenda and at the bottom of the Spanish one; the second goal: *supplying municipality information via telecommunications networks (internet, teletex, telephone and so on)* seems to be in the priority list of Austria and far from the objectives of the Spanish politicians; *promoting municipality services via telecommunication networks* appears to be a key aspect for both France and Austria and far again from the immediate goals fo the Spanish local governments. And last but not least, using ICTs in transport planning (intelligent road signs, online traffic information, etc.) fits mainly the politics of France and keeps a poor profile within the Spanish applied local policies.

4. City networking: from processes to policies

It is commonly assumed that, historically, cities have always existed in environments of linkages, both material flows and information transfers. They have acted as centers from where their hinterlands are serviced and connected to wider realms. Cities have played an important role in the process of globalization as centers for information exchange and specifically one of the greatest ironies of the age of globalization has been the continued prosperity and growth of cities. Currently, modern technology, political and economic evolution have intensified the systems of relations and most importantly they have changed their quality. Analyzing the transformations of systems of cities some authors point to the influence of history – known in the cluster literature as path-dependence – as well as to the question of specialization vs. diversity. Connected with this last point is the agglomeration of knowledge workers in areas blessed by a favorable mix of venture capital firms, universities and research institutes, and a crackling atmosphere.

Groups of firms, linked in a dense web of contractual agreements and partnerships, can thus capitalize on the coordination of specialized local industrial clusters. Without the capabilities of new information and communication technology, these networks would be unmanageable. "What challenges await urban planners as they grapple with the implications of these new technologies? How can cities plan to take advantage of the growth of interurban communications or mitigate its negative consequences?" (Townsend, 2001b: 1711). The answer to this questions is offered by the same author: urban planners must address immediately three primary areas of concern: dependency on other cities and urban areas, accessibility to global Internet backbone networks, and proficiency with communications technology. "Cities that are plugged in to new communication networks face a very different set of questions dealing with the antithesis of competitiveness, dependency … In fact, it is likely that the cities with the most dependencies have and will continue to thrive the most … It is clear that urban planners and policy makers must now develop an increasingly complex conceptual model of their region's telemediated connections to the rest of the world" (Townsend, 2001: 1712). Moreover, "it is clear that a divide is emerging between cities and regions that are well connected to the global Internet, and those that are not … There is a compelling argument for public subsidy and even operation of backbone links from disconnected cities to international backbone hubs". Finally, "the virtual city of Web sites, emails, and intellectual property is just as important as the physical city of streets, bridges, and sewers.

Cities need to invest in developing the navigational tools and workforce skills needed to traverse this new urban landscape" (Townsend, 2001:1713).

Figure 2. Network activities in the field of ICTs by city size (% over the total sample, even the percentages exceed 100% due to the possibility of chossing more than an option)

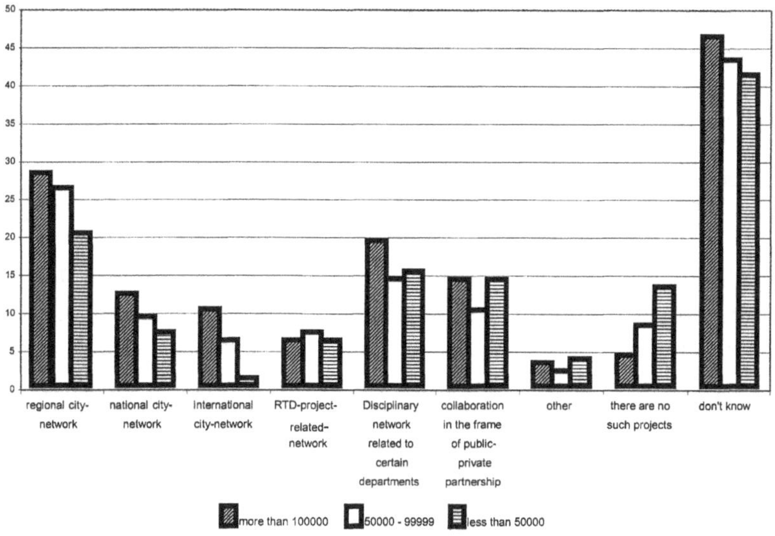

a. A start-point: co-operation and competition

Changing the scope and scale of analysis, we wondered if: *is there a formal strategy plan/action plan/program on ICTs or the Information Society at national or regional level?* Unfortunately, the results justify the lack of consistency as far as the ICTs policy is concerned. The percentage of "don't know" answers (between 64% and 71%) confirms the information gap between the local actors and regional/national policies. The knowledge about national/regional level policies is the lowest and not surprisingly then are the notions on their influence the vaguer the more specific the questions about potential areas of influence were phrased.. Ireland and France seem to have benefited mainly from national level plans/programs, again France and Spain from regional ones while Norway is over-represented in the group of negative answers and Germany confirms the information gap between the local actors and regional/national policies. Anyway, only France excess the 50% of answers identifying national or regional strategic plans in its own cities.

Bearing in mind this knowledge "black-hole", one step further in order to reveal the influence of national/regional plans upon the local policies: *does this national/regional plan have any influence on the activities of your municipality?* uncovers that actors living in little towns attach certain influence on *organisation/co-ordination of local ICT activities* followed by the *financial support*, while medium cities highlight again *financial support*, *research and development* and *legal support.* Anyway we must remember that only a few respondents were able to state an opinion as to whether such regional or national documents has any influence on their municipalities' ICT activities. Really, is the financial support under-estimated? If we trust the statements of different authors, government and other institutional funding is overwhelmingly of a pump-priming nature. The problem for any local community initiative is how to sustain an initiative after the initial funding ends. In one sense, the success or lack of it of an initiative can be judged by looking at what happens once the funding ends. Norway's actors attach it a great influence, just the contrary the Dutch actors do, while in this case the Spaniards are the ones who fit in into the category of "I don't know". As far as the scope of this influence is concerned Ireland highlighted both the *organisation/co-ordination of local ICT activities* , followed by the *financial support* , Germany is over-represented in the category of *research and development* and Austria in that of *legal support.*

When we directly ask the local actors about the existence of ICT projects in their municipalities: *is your municipality currently running any projects (e.g. RDT-projects, pilot studies) on ICTs?* The answers disclose the current deep engagement level of medium cities (54.3%), which contrast with the low profile of little towns and the knowledge gap the local actors from big cities (42% don't know) manifest. By testing these answers the idea that in most medium-sized cities, the public sector does not play a sufficiently active part in encouraging the use of ICTs is in question. In any case, we still must accept that the introduction and deployment of information and communication technologies largely depend on personal initiative and enterprises.

The answers disclose the current deep engagement level of Norway and France and the future plans of Irish municipalities, which contrast with the low profile of Austria municipalities and the knowledge "black-hole" the German local actors manifest.

The financing of the previously mentioned projects on ICTs is a good point to verify the engagement level of the local/regional/national administration or their dependency on European/international funds not to mention the level of public-private partnership. Leaving aside that in all the cases municipality budgets are the main financing source, not surprisingly are big cities the most benefited from the European Commission while medium cities score higher in national/regional funds and little towns cover the financing using their own municipality budgets as well partnership with the private sector. Not surprisingly are Ireland and Spain the most dependant on European Commission and other international funds, both are European countries supported by cohesion and structural funds due to their relative underdevelopment in European context. National funds are well represented in the Netherlands, regional ones in Spain, and the municipality budget cover generously France and Norway projects. As far as partnership with the private sector is concerned, the Netherlands reaches the highest involvement level.

Still talking about the ICT-projects, the question focusing on: *what department are/were involved in these projects?* allows a multiple choice response. Our informants highlight the *urban planning and urban development department* in the cases of medium cities and little towns, medium cities again score over the average for the involvement of *economic promotion department* and *transport* one while little towns seem to concentrate on *communication, environment* and *social affairs/education department.* Big cities, even without keeping a low score profile in all the categories, stress the engagement of *economic promotion departments*, something understandable if we remember that most interviewees regard ICT mainly as an economic potential and location factor. Our interviewees highlight the *urban planning and urban development department* in the cases of Austria and Norway, *economic promotion department* clearly leads the case of France, while *communication* is well represented both in the cases of France and Norway, Spain seems to concentrate on *environment* and *transport*, and *social affairs/education department* is again the domain of Norway municipalities. The low involvement of the transport department could be understood? Anyhow, municipal governments and urban development agencies look to be currently engaged in a race to cope with the radical transformations which are underway in the economic makeup and functioning of modern cities. Telecommunications and telematics are now involved as key "generic technologies" in attempts to innovate and compete in all stages of the production process and all economic sectors. Key results are new telematics-based ways of organizing and covering space and territory and new diversified prospects for cities as centers of economic activity.

b. Specific policies to impel networking

Although the orientation and goals of interurban networks vary, they all embody a recognition of interdependence, a pooling of resources, an exchange of information, a development of trust, a mutual orientation, and a long-term collaborative commitment between cities (Stoker and Young, 1993). This policy innovation falls into two broad areas: the development of intraregional, national, and transborder networks between local groups of collaborating cities, and the development of international interurban telematics networks.

Intra-regional, national, and trans-border telematics-based urban networking centres on the goals of economic cooperation and information exchange. Such networks are most highly developed in France, where telematics are being centrally addressed within mainstream urban and regional planning activities. French municipalities have unusually strong institutional linkages with telecommunications policymakers through the URBA 2000 initiative and the "Observatoire des Télécommunications dans la Ville" – a body which brings urban, state, and telecommunications representatives together to support urban and interurban telematics applications (Graham, 1995). The extreme municipal fragmentation of French local government presents a particularly strong demand for the emerging ethos of "intercommunality" which is becoming embodied in the use of intercommunal telematics networks. The first national expression of this trend was "Le Réseau Villes Moyennes" (RVM: the medium-sized town network), set up in 1987. Linking up towns with populations of between 20,000 and 100,000, and based on the ubiquitous Minitel viewdata system, this service incorporates a professional electronic messaging service enabling the exchange of up-to-date questions and answers and a directory of "best-practice" project profiles (Graham, 1995).

This is part of a broader shift toward inter-communal and inter-tier strategic planning that has emerged in France since the decentralisation reforms of the early 1980s. These "new solidarities" between cities are based on the identification of what some development agencies call "complementary poles of excellence" in the different city economies (Lavocat, 1989).

Following a similar path, even at the less ambitious regional scale, the "Infoville" initiative, launched by Spain's local Valencia Government started at the summer of 1995 aiming at fostering and facilitating transition of the Valencia Autonomous Community to the Information Society. In order to achieve this goal, they have integrated several cities into the "InfoVille Ciudades Digitales" Network (Villena, Torrevieja, Catarroja, Vall d'Uuixo, Burriana, Oliva and Altea). All of these cities are of a population range between 15,000 and 30,000 inhabitants. The original idea was focused on different fields covering education and health system as well as the integration of the various layers of public administration. The creation of the parallel "InfoVille Organizaciones Digitales" Network began with "InfoCampus" at the University of Elche where 1,500 students of this new university, apart from having access to the services in "Digital cities", can reach other specifically university services.

Interurban telematic projects are also being explored as ways of improving the collaboration and integration between adjacent border areas within the EU. The economic integration of adjacent border areas in Europe has long been heavily restrained by the costs and delays caused by the continuation of nationally centralised postal and telecommunication systems. In certain circumstances, telematic networks are being developed to help break down these artificial barriers by linking adjacent border regions, a policy objective that is being heavily supported by the EU's INTERREG programme. Such projects can drastically cut the costs and improve the speed of interactions between adjacent firms and local authorities. In the border area between Germany and the Netherlands, for example, eighty urban and rural public authorities are trying to use an electronic mail service to overcome the problems caused by national boundaries as part of the "CEC' EUREGIO" project (Graham, 1995). As with the RVM, the emphasis is on exchange of best-practice policy lessons via electronic messaging.

International interurban telematic networking initiatives are focused as much on improving the lobbying for national and European resources as for information exchange and the stimulation of economic collaboration. A major stimulant to the development of these networks has been the EU's own initiatives. The fact that many of the structural funds that support economic, social, and environmental projects in cities require interurban collaboration across national boundaries has added an important impetus to the wider forces promoting the search for collaborative networks. Sometimes, the EU provides even more direct support to a wide range of emerging collaborative strategies based on interurban networks which are geared towards trying to foster trans-European economic, institutional, and infrastructures integration (McGowan, 1993). These trends are parts of a wider shift towards the "Europeanisation" of local government, as they come to deal more and more directly with European institutions and foreign local policy-makers (Goldsmith, 1993).

Telematic applications are increasingly being explored as ways of improving the effectiveness and efficiency of trans-European interurban networks. Three emerging examples can be highlighted: the Eurocities/Telecities network, the European Urban Observatory (EUO), and the European Regions's Network for the Application of Communications Technology (ERNACT) initiative.

Eurocities project was born as an informal club representing about sixty "second cities" from across the European Communities (Parkinson, 1992). The activities of the network are diverse, ranging from lobbying the EU, influencing EU policy decisions, exchanging best practices, and supporting economic links (Dawson, 1992). The Telecities initiative aims to "promote and develop telematics initiatives which support the regeneration or urban areas through economic development and strategies to tackle unemployment, social development and improved quality of life, new solutions to combat social exclusion, (and) maximising the resources available to cities in order to support regeneration projects" (Telecities, 1994). To complement the broader Telecities network within the EU, electronic networking is also being developed along an East-West axis via established twin-cities arrangements. Once again, the EU is directly supporting such East-West interurban networking through the Ouverture initiative (Graham, 1995). Currently (February 2002), the Telecities initiative has enlarged its city network to include 123 members: more than 100 European cities, 2 public administrations and 7 business members.

The European Urban Observatory stems from an EU study to explore the role of science and technology in the future of European cities. A key conclusion of that study was that real interurban information systems were required to improve the quality of interurban comparative policy research. The EUO, a project funded through the RECITE (Regions and Cities in Europe) programme, is an attempt to address these issues. The EUO provides a forum for the exchange of information of common interest, best practice in addressing problems to do with economic, social, and environmental development, and to facilitate technology transfer.

The third example is the ERNACT programme. This project puts together the United Kingdom, the Republic of Ireland, Denmark, the Netherlands, and Belgium "to improve the ways in which regional and local authorities apply communications technology in the performance of their duties in order to enhance the development of their regions" (Graham, 1995).

c. An overview on city networking according to the results of the TCV survey

City networking, a key element of the new institutional infrastructure and a convention of the new neo-liberal practices is assessed in the following question: *in which network activities in the field of ICTs, i.e. co-operation with other actors, is your city involved?* This multiple choice question reveals that big cities are the most active nodes within the information network at a national as well as at international level. In fact they score over the average in *regional city-networks, national city-networks, international city-networks, disciplinary network related to certain departments* and *collaboration in the frame of public-private partnership*. Even if they complain about the lack of both systematic and specific ICT policies, this fact does not appear as to hindering their capacity to engage in more or less complex city networking strategies. Medium cities only excel in *RDT-project-related network*, while the category of *there are no such projects* is dominated by little towns. Here again we must warn about the knowledge gap disclosed among our interviewees, the average of "don't know" answers reaching 44.7%; there appears to be not much real knowledge on municipal ICT networks. Taken the vagueness of these answers into account, one does not get a clear picture whether the cities are indeed networking or whether respondents believe they are. All sorts of inter-city co-operation and association programs now exist, from town twinning at the lowest level to various European Union initiatives aimed at especially created networks of European cities. Or, in another words, the role of networking in facilitating the creation of the "European village" is what is at stake.

Switching the view point to the country scale (Table 1), the previous multiple choice question reveals that *regional city-networks* are the privileged scale for France and the Netherlands, while Austria is over-represented on *national city-networks*, and France again excels on *international city-networks*. Austria appears as the "champion" of *RDT-project-related network, disciplinary network related to certain departments* and *collaboration in the frame of public-private partnership*. The category of *there are no such projects* is dominated by Ireland and that of *don't know* by Germany. Cities, though they must compete individually, can also compete better if they gather together in networks. These networks are particularly prevalent in Europe, since "to participate in Europe, at whatever level, is increasingly a matter of being "on the network" (Barry, 1996).

Table 1. National city allocation according to city networking policies

	Austria		France		Germany		Ireland		Netherlands		Norway		Spain	
	Total	%	Total	%	Total	%	Total	%	Total	%	Total	%	Total	%
regional city-network	24	26.4	37	32.5	218	27.4	13	16.7	41	32.0	30	16.6	46	26.7
national city-network	21	23.1	24	21.1	62	7.8	7	9.1	21	16.4	11	6.1	21	12.2
international city-network	11	12.1	21	18.4	46	5.8	9	11.7	11	8.6	4	2.2	20	11.2
RTD-project-related-network	10	11.0	7	6.1	49	6.2	1	1.3	6	4.7	12	6.6	15	8.7
Disciplinary network related to certain departments	27	29.7	6	5.3	146	18.4	6	7.8	18	14.1	25	13.8	50	29.1
collaboration in the frame of public-private partnership	17	18.7	11	9.6	112	14.1	6	7.8	10	7.8	33	18.2	24	14.0
other			10	8.8	19	2.4	1	1.3	4	3.1	5	2.8	5	2.9
there are no such projects	9	9.9	2	1.8	26	3.3	11	14.3	6	4.7	23	12.7	13	7.6
don't know	29	31.9	30	26.3	419	52.7	32	41.6	46	35.9	80	44.2	61	35.5
Total	91	100.0	114	100.0	795	100.0	77	100.0	128	100.0	181	100.0	172	100.0

d. City networking in the context of urban hierarchies evolution

The network city operates in an economy where the transport costs of the highest value products –information and knowledge- are fairly insensitive to distance. Furthermore, the widespread proliferation of sophisticated information technology has undermined the need to coordinate operations in a central headquarter. Both trends, as well as emerging patterns in air transportation, point towards a continuation of this distributed pattern of urbanization in the future –a structure of urban activity nodes connected by many redundant pathways. The real challenge to information development is not physical or financial capital, but human capital. Places that already have a large skilled force will become even more special and attractive to footloose firms. Quick-fix policies that try to use physical infrastructure as an incentive will fail without substantial commitments to develop the skilled labor force which is necessary to make these facilities a valuable asset (Townsend, 2001). A new type of global city, the network city is emerging. These highly prosperous regions are attracting the skilled workers and the investment in infrastructure that is needed to sustain their growth, often at the cost of lost opportunities for other regions, and they are being organized into an exclusive and highly codependent economic system.

Thanks to ICTs, the cities, doesn't matter their size, play a role as interface between the geo-space and the cyberspace (following Castell's terminology) being a part of it depending on their connection to the telematic networks. Because of this more free way for constructing the urban networks. The potentialities of cities could be taken into account in the new world networks; this is particularly important for middle and small cities, that would be able to break down their previous subordinate position to the big ones into the former hierarchical traditional distance-based networks (Valenzuela, 2000: 28-30). Even if historically the idea of urban hierarchy was accepted without discussion, having in mind the new urban and technological processes, it needs new empirical grounding and theoretical interpretation (Taylor, 1997: 323).

Nevertheless, the present fashionable global outlook has itself many roots. It results from an inheritance from the past, systematic expansion of long-distance relationships throughout the known and accessible space. It also results from the technological, economic and cultural evolution which recently improved the accessibility of diverse parts of the world in spectacular manner. The reasons for the intensification of external relations seems easy to understand: modern technology offers the means for moving people, information, goods and money easily and quickly; moreover recent political and economic evolution has removed or made more permeable the old barriers to the international flows of people, money and information, and to some extent to the movement of goods (Gottmann, 1991). Anyway, if we trust the thesis of some authors the longer a city has had trading, political and/or cultural functions, the more established and involved its networks of relations will be, and the more lasting its role as a partner in the networks. What actually has been happening looks like a proliferation of new structures, many of which are not yet called cities but which gradually acquire an urban semi-centrality, while older cities and towns have shown on the whole amazing resilience and capacity to adapt and develop their networks.

5. Concluding remarks: starting hypothesis and findings

In order to verify one of the hypotheses of the project we would need to evaluate what constitutes an over- or under-estimation of ICT impact which would require objective indicators about the actual impact of the "Information Society" on urban development. Lacking such indicators as well as a definition and means of measurement of what are suitable strategies, our first results can merely be assessed in terms of individual perceived impact, of levels of strategic incorporation to the plans and practices sphere and their relationships. That is to say, answers of the interviewees are subjective sort of ways to approach reality.

In this sense, we find out that in general terms local actors rarely doubt that Information Society does have an impact on urban development if unevenly distributed in all areas, they attach a great influence to ICTs for those aspects linked to "urban importance"/urban economies, mainly for big cities, competition with other cities and the potential to attract service companies, or in another words the process related with tertiary activities and urban marketing. The capacity of ICTs to attract new residents and control the people mobility and especially to lead physical and social trends like sub-urbanisation and socio-spatial segregation appears clearly underestimated.

Belief patterns slightly vary depending on city size criteria: Starting from an over-estimated self image big cities tend to stress future increasing competition (economic/urban marketing trends) processes among cities, along with a positive valuation of ICTs in the competitive position of the city as well as support for urban networking practices. Opposite to this economic/locational role of ICTs they skeptical valuate the ICTs potential to increase citizen participation and democracy. The strategies displayed following the responses of this group of actors focus mainly on administrative and infrastructural dimension goals, they have poor systematic and specific ICT policies and they manifest a deep knowledge gap about the existence of plans at regional/national level. However they have succeeded to get a meaningful percentage of ICT projects financed by the European Commission and they display a rich and diversified city networking experience at geographical, thematic and public-private partnership scale and scope. As for the relationship between impact assessment and strategies, we find a certain coherence within the frame of urban neo-liberal economies stressing competitiveness and selected links or communication paths, and neglecting systematic and specific ICT policies to concentrate in targeted ICT projects.

Medium cities depict a quite contradictory belief pattern combining increasing co-operation among cities (and high valuation of Internet and email) and a soft "dystopian" low enthusiastic valuation of ICTs to increase the competitive position of the city within the European urban network. They show the same deep knowledge gap about the existence of plans at regional/national level, and they concentrate on specific ICT policies and ICT municipal projects. Apart from their own budget they benefit from a wide range of financial support: regional/national ICT plan, regional government funds, national funds, etc. and they score over the average in city networking activities dealing with RDT-project related-network. The comparison between impact assessment and strategies uncovers a quite ambiguous approach to the potential of ICTs to change city role and to enforce its competitive position, at the same time they have great expectations of increasing city co-operation not fitted by city networking strategies. They lack systematic ICT policies and concentrate on specific ones as well as on ICT municipal projects.

Last but not least, little towns yielded the most humble self-valuation about their capacity to be labeled as ICT and/or multimedia center, even though they pointed to future trends of increasing competition processes among cities, along with a positive valuation of ICTs in the competitive position of the city. They also have a high awareness level of the ICT potential for improving administrative and participatory dimensions while the idea that ICTs will support urban networking with another European cities is reluctantly assessed. The strategies that parallel these beliefs reinforce the role of communication and social affairs/education departments in ICT projects. As far as strategic comprehensiveness is concerned they show the lowest profile within the sample, their answer to the existence of ICT policies is extremely polarized and contradictory. Their financial support come from national/regional plan (if that exists), municipality budget and partnership with the private sector. The relationship between impact assessment and strategies seems to enhance the ICT possibilities for socio-economic integration as well as general participatory and democratic dimensions. Anyway, it must be remembered the clue presence in this category of Norwegian cities whose strong welfare state culture could have slightly biased the sample. In brief, and following the conclusions of the Austrian team about the results gained from the quantitative survey "while we cannot indicate whether impact assessments over- or under-estimate the influence of information societies or whether strategies are appropriate or not, we do find that impact beliefs and strategic comprehensiveness apparently tend to go hand in hand".

Nevertheless, having in mind the great number of actors leading the introduction of ICTs, and their influence on the processes of network city and city networking, as well as the fact that private actors are often understood as the key characters in deploying urban information societies, it must be wondered if municipal and local involvement is both critical and realistic. Following the thesis of Graham "care must be taken to avoid being seduced by the luster and hype surrounding cyberspace and virtual cities. I have doubts about the extent to which convivial, face-to-face interaction and the public, democratic realm of urban places can be genuinely substituted by interaction virtual cities. The capabilities for mediating social and economic activities via telecommunications are often widely exaggerated; the continued importance of concentrations of activities in urban places is often underrated in the clamour to hype-up the "telecommunications revolution"...The message, then is that careful policy innovation is certainly important but that this needs to be geared very directly to stimulating improvements in real cities...In short, urban places and electronic spaces need to be planned in parallel" (Graham, 1995:201).

On the other side, when considering the basic tendencies transforming systems of cities and city networking processes we must realize that apart from public ICT policies a whole set of variables explain city potential for changing its role, as well as for creating or enlarging a complex urban network. One of the findings of our research has been that politicians have to learn that the social shaping of ICT has to be supported. ICT-policy should be understood as cultural and social policy. Until today there is no clear view how the benefits of ICT could fit into a theory of modernisation of the civil society itself. On the contrary, the rhetoric of ICT seem to reflect the ideology of modernization and ICT policies have become a pure fetish.

References

Allen, J., Massey, D. and Pryke, M. (eds.) (1999): *Understanding Cities: Movement and Settlement*, Milton Keynes: Open University Press.

Amin, A, and Graham, S. (1999): "Cities of connection and disconnection" (in) Allen, J., Massey, D. and Pryke, M. (eds.): *Understanding Cities: Movement and Settlement*, pp. 7-48, Milton Keynes: Open University Press.

Amin, A. and Thrift, N. (1992): "Neo-Marshallian nodes in global networks", *International Journal of Urban and Regional Research*, 16, pp. 571-587.

Barry, A. (1996): "The European Network", *New Formations*, 29, pp. 26-38.

Beaverstock, J.V.; Smith, R.G. and Taylor, P.J. (2000): "World-city network: a new metageography?", *Annals of the Association of American Geographers, 90, 1*, pp. 123-134.

Berry, B.J.L. (1996): "Technology-sensitive Urban Typology", *Urban Geography, 17, 8*, pp. 674-689.

BIS (1998) *TCV-TELECITYVISION. Information society and urban development in European Comparison. Technical annex.* Berlin, 16 p.

Braczyk, H.J.; Fuchs, G. and Wolf, H.G. (eds.) (1999): *Multimedia and Regional Economic Restructuring*, London, Routledge.

Bratzel, S. and Dienel, L. (1999) *National Urban Systems and ICT in Europe. Some comparative Remarks.* Berlin, Technical University Berlin (Centre for Technology and Society) 39 p.

Camagni, R. (1993): "Organisation économique et réseaux de villes", (in) Sallez, A. (ed.): *Les Villes, Lieux Europe*, Paris: Datar, pp. 107-128.

Dawson, J. (1992): "European city networks: experiments in transnational urban collaboration", *The Planner*, 10, January, pp. 7-9.

Dematteis, G. (1994): "Global networks, local cities", *Flux*, 15 (January-March), pp. 17-24.

Gillespie, A. and Williams, H. (1998): "Telecommunications and the reconstruction of regional comparative advantage", *Environment and Planning A*, 20, pp. 1311-1321.

Goldsmith, M. (1993): "The Europeanisation of local government", *Urban Studies*, 30, pp. 683-699.

Gotmann, J. (1991): "The dynamics of city networks in an expanding world", *Ekistics*, vol. 350-351, pp. 277-281.

Graham, S. (1994a) "Networking cities: Telematics in urban policy - a critical review", *International Journal of Urban and Regional Research*, 18 (3), 418-432.

Graham, S. (1994b), "From urban competition to urban collaboration? The development of inter-urban telematics networks", *Environment and Planning C: Government and Policy*, 13, 503-524.

Graham, S. (1995): "Cyberspace and the city", *Town and Country Planning*, 64 (8), pp. 198-201.

Graham, S. (1997): "Cities in the real-time age: the paradigm challenge of telecommunications to the conception and planning of urban space", *Environment and Planning A*, 29, pp. 105-127.

Graham, S. and Marvin, S. (1996): *Telecommunications and the city: electronic spaces, urban places*, London and New York: Routledge, 439 p.

Graham, S. and Marvin, S. (2001): *Splintering Urbanism. Networked infrastructures, Technological Mobilities, and the Urban Condition*, London and New York: Routledge.

Harvey, D. (1989): "From managerialism to entrepreneurialism: the transformation of urban governance in late capitalism", *Geografiska Annaler B*, 71, pp. 3-17.

Lavocat, E. (1989): "From networked towns to towns networks: new solidarities, new territories and the impact of telecommunications", *Netcom*, 2 (2), pp. 381-403.

Leitner, H. (1990): "Cities in pursuit of economic growth: the local state as entrepreneur", *Political Geography Quarterly*, 9 (2), pp. 146-170.

Leitner, H. and Gardener, M. (1993): "The limits of local initiatives: a reassessment of urban entrepreneurialism for urban development", *Urban Geography*, 14 (1), p. 57-77.

Leitner, H. and Sheppard (1998): "Economic uncertainty, interurban competition and the efficacy of Entrepreneurialism" (in) Hall, T. and Hubbard, P.: *The Entrepreneurial city*, Chichester: John Wiley, pp. 285-309.

Lovering, J. (1995): "Creating discourses rather than jobs: the crisis in the cities and the transition fantasies of intellectuals and policy makers", (in) Healey, P., Cameron, S., Davoudi, S., Graham, S. and Madani-Pour, A. (eds.): *Urban Management*, London: Belhaven Press, pp. 109-126.

McGowan, F. (1993): "Trans-European networks: utilities as infrastructures", *Utilities Policy*, 3 (July), pp. 109-185.

Mitchell, W.J. (1999): *E-topia: "Urban life, Jim – but not as we know it"*, Cambridge MA: MIT Press, 184 p.

Mitchelson, R.L. and Wheeler, J.O. (1994): "The flow of information in a Global Economy: The role of the American Urban System in 1990", *Annals of the Association of American Geographers*, 84 (1), pp. 87-107.

Moss, M.L. (1996): *Telecommunications Policy and Cities*, Taub Urban Research Center, New York University, http://urban.nyu.edu/research/telecom-policy-cities/

Moss, M.L. and Townsend, A.M. (1999): *How Telecommunications Systems are Transforming Urban Spaces*, Taub Urban Research Centre, New York University, http://urban.nyu.edu/research/telecom-urban-spaces/

Paddison, R. (1993): "City marketing, image reconstruction and urban regeneration", *Urban Studies*, 30, pp. 339-350.

Parkinson, M. (1992): "City links", *Town and Country Planning*, 61 (September), pp. 235-236.

Petrella, R. (1995): "A Global Agora vs. Gated City Regions", *New Perspectives Quarterly*, winter: pp. 21-22.

Robson, B. (1992): "Competing and collaborating through urban networks", *Town and Country Planning*, (September), pp. 236-238.

Smith, D. and Timberlake, M. (1995): "Cities in global matrices: toward mapping the world system's city-system" (in) P. Knox and P. Taylor (eds.): *World Cities in a World System*, pp. 67-68, Cambridge: Cambridge University Press.

Stoker, G. and Young, S. (1993): *Cities in the 1990s*, Harlow, Essex: Longman.

Swyngedow, E. (1989): "The heart of the place: the resurrection of locality in an age of hyperspace", *Geografisker Annaler B*, 71 (1), pp. 31-42.

Taylor, P.J. (1997): "Hierarchical tendencies amongst world cities: a global research proposal", *Cities*, vol. 14, 6, pp. 323-332.

TELECITIES (1994): "Telecities: the European telematics partnership", promotional brochure.

Townsend, A. (2001a): "The Internet and the rise of the new network cities, 1969-1999", *Environment and Planning B: Planning and Design, 28, 1*, pp. 39-58.

Townsend, A. (2001b): "Network Cities and the Global Structure of the Internet", *American Behavioural Scientist, 44, 10*, pp. 1697-1716.

Thrift, N. (1999): "Cities and economic change: global governance?", (in) J. Allen, D. Massey and M. Pryke (eds.): *Understanding Cities: Movement and Settlement*, pp. 271-319, Milton Keynes: Open University Press.

Valenzuela, M. (2000): "Les réseaux urbains, un concept en voie de revision. Une approche de leur théorie et de leur pratique en Espagne", *Géographes Associés*, 24, 200, pp. 27-37.

Veltz, P. (2000): "European cities in the world economy" (in) A. Bagnasco and P. LeGalès (eds.): *Cities in Contemporary Europe*, Cambridge: Cambridge University Press, pp. 33-47.

Zook, M.A. (2001): "Old hierarchies or new networks centrality?", *American Behavioural Scientist*, 44, 10, pp. 1679-1696.

Annex

A1: City size: More than 100,000 Inhabitants

	Meaning: Perception of needs/possibilities	Equipment and infrastructure at work	Doing: Plans and policies
City self image/ urban marketing	- Over-estimated self image - Increasing competition processes among cities - Positive valuation of ICTs in the competitive position of the city within the European urban network		
Administrative dimension			Promoting administrative dimension goals
Partizipatory dimension	Sceptical valuation of ICTs to increase citizen participation		
Infrastructural dimension		Lan/man connections	Promoting infrastructural dimension goals
Planning dimension			- Lack of systematic and specific ICT policies - Ignorance of Plans at regional/national level
City networking	ICTs will support urban networking with another European cities		- Regional city-network - National city-network - International city-network - Disciplinary network related to certain departments - Collaboration in the frame of public-private partnership
Economic dimension	- Increasing competition processes among cities - Sceptical valuation of ICTs in the competitive position of the city within the European urban network		ICT projects financed by European Commission

A2: City size: Between 50,000-99,999 inhabitants

	Meaning: Perception of needs/possibilities	Equipment and infrastructure at work	Doing: Plans and policies
City self image/ urban marketing	- Increasing co-operation among cities - Less enthusiastic valuation of ICTs to increase the competitive position of the city within the European urban network		
Administrative dimension		High frequency of use of: - word processing - email - spreadsheet	
Partizipatory dimension			
Infrastructural dimension	High valuation of Internet and email	- High percentage of PC users - Modem	
Planning dimension			- Specific ICT policies - ICT municipal projects - Ignorance of Plans at regional/national level
City networking			RDT-project related-network
Economic dimension	Less enthusiastic valuation of ICTs to increase the competitive position of the city within the European urban network		Financial support of regional/national ICT plan Financing of ICT projects: - Other international funds - National funds - Regional government funds Economic promotion department involved in ICT projects

A3: City size: Less than 50,000 inhabitants

	Meaning: Perception of needs/possibilities	Equipment and infrastructure at work	Doing: Plans and policiesS
City self image/ urban marketing	- Lack of ICT and/or multi-media centres - Increasing competition among cities - Positive valuation of ICTs in the competitive position of the city within the European urban network		
Administrative dimension	ICTs improve communication within city administration	High frequency of use of PC, Internet, intra-net and databases	Communication depart-ment involved in ICT projects
Partizipatory dimension	ICTs increase citizen partici-pation		Social affairs/education department involved in ICT projects
Infrastructural dimension	High valuation of cellular phone	Isdn/asdl Cables	
Planning dimension			- Polarized answer to the existence of ICT policies - Ignorance of formal strategy plans at na-tional or regional level
City networking	Shy reluctance to the idea that ICTs will support urban net-working with another European cities		Poor city networking
Economic dimension	- Increasing competition among cities - Positive valuation of ICTs in the competitive position of the city within the European urban network		- Financial support of national/regional plan ICT local projects fi-nanced by: - Municipality budget - Partnership with the private sector

8 Perceptions of ICT and its Impacts on City Competition

Leon Hempel, Hans-Liudger Dienel

Abstract

The impact of the increasing use of Information and Communication Technologies (ICTs) on the city has been the topic of the comparative European research project TeleCityVision (TCV)[1]. Given the major change of assessing the impacts of ICTs since the demise of the New Economy in Spring of 2000 the following paper firstly provides a brief introduction into the subject and how it has changed within the last decade. Secondly, it reconsiders the research approach of the project in order to set a frame to present the outcomes of the study. Thirdly, it presents the results of the work: The perception of ICTs is discussed within the contexts of rising city competition.

Introduction

Today urban officials are under constant political pressure to develop all kinds of innovative concepts for compensating urban stagnation. The permanent crisis of the city since the 1970s has only been managed, but not solved. In the beginning of the 1990s however a new solution has been envisaged in the form of ICTs. In this respect the new technologies were understood as means to solve the common post-industrial crises of cities in regard to diverse urban problems,

[1] TCV Project financed by EC, co-ordinated by BIS, participants: BIS (Germany), COMTEC (Ireland), ESI (Netherlands), ETS-UAM (Spain), ICCR (Austria), STS (Norway), ZTG (Germany)

such as budget, unemployment, transportation, population, but also the question of the city's self-representation. Today the medial image of the urban area serves to present the city to the world. For example, the so-called Bangemann Report of 1994, advertising a "New World Order for Global Communications" on behalf of the European Commission, stated: "What is required is that the Information Society becomes a key driver for all other policies, for instance job creation." (Bangemann 1997) As a panacea ICT has led to a flood of concepts during the 1990s stimulating new spirits for an otherwise sombre urban future (see Melody 1996: 243). The wave of new communication technologies should wash away the old foundations of the industrial cities and working societies. We were all suppose to witness a spectacular urban evolution of an unknown dimension in which interaction and communication would be the most prominent features of the new global urban system.

Especially the overwhelming success of the internet as a kind of huge information marketplace inspired a whole range of visions of a new urban future. Temporal and spatial barriers tumbled with the implementation of the 'world wide web'. Advantages and disadvantages of spatial situations of cities suddenly looked as if they would become irrelevant. What once begun as a local concentration of marketplaces in cities should now change to an unlimited market with world-wide access for both providers and consumers. Many commentators saw a big chance for the cities in this. They argued that the cyberspace and the new economy would not only diminish the significance of urban centres but would offer new challenges for every city in the world to participate in the global marketplace.

In the middle of the 1990 conceptions of urban restructuring and revitalisation were just about to reach their limits. The discovery of the virtual space opened new space of unlimited opportunities. The annulment of space and time would not only alter forms and functions of our cities but also change urban life. Communication would improve. Knowledge and information therefore should become available to all city members. ICT would enable higher degrees of networking and better co-operation throughout the world and also within the cities. Thus, the making of urban politics should change and public participation should be facilitated. Moreover, benefits in education and qualification could be expected as everybody would now have access to it. Not to forget, administrative work should become more efficient and transparent. Leggewie and Maar have argued in 1998 that the internet should not be seen as a platform for marketing and economics, but as a medium to improve the shaping of public opinion in order to revitalise and legitimate democratic political processes.

Given this utopian potential of ICT, it must be pointed out, that also threats have been seen by the challenges of ICT right from the beginning. The information society, as Manuel Castells argued, would change the structure of our cities tremendously: The so called "informational city" tends to be a "dual city" where new social conflicts will arise between "high value making groups" on the one hand and "devalued social groups" on the other hand (Castells 1989). In opposition to the common "space of places" Castells has seen a "space of flows" arise which networks are spread all over the world, but taking a closer look are also centralised in the manifold economic centres of the cities. Both types of space get into conflict, he expected. The very dynamic abstract space of information would break the material space with its everlasting problems and would deepen segregation and accentuates inequality. Castells therefore asked how both spaces can be associated again in terms of a socially balanced relationship. (Castells 2001). Saskia Sassen argued similar, when she pointed out that the rise of the global economies are followed by new migration flows. The global city would not only attract economic power but also people from all over the world. Work power, productivity, efficient socio-cultural and educational facilities should still be important conditions in order to enable the economic power of the global city. But large parts of the inhabitants, she assumed, that are needed to keep the business going, will live socially marginalised (Sassen 2000).

Other threats have been seen. Already by the end of the 1980s Oscar Gandy mentioned that there is a close relationship between ICT and surveillance technologies. He reported "that the real source of growth in both the information work force and the development of information technologies is not to be found in any transformed consumer demand, but in the continually expanding surveillance requirements of multinational corporate enterprises. Indeed, for some observers, 'information society' is a misnomer that hides the extend to which industrial societies have in fact become surveillance societies" (Gandy 1989: 61). ICT would connect people throughout the world, they would provide new forms of communication, offer access to unforeseen sources of information and allow all actors, from individuals to cities, an infinite space to reach distant awareness of others. But at the same time ICT would provide the basic technique for surveillance tools. These could counteract effects of the information society such as the assumed revitalisation of democratic legitimisation.

In the meantime we know that the expected effects have not occurred as it has been wished first. The demise of the so-called New Economy in the Spring of 2000 has soberly underscored the fact that the tertiary sector can hardly compensate for the loss of industrial workplaces in the long run. First of all managers of the financial sector needed the new technologies for their businesses. The private sector was the driving force within the process towards the information society. After the collapse however the visionary power of conceptions have calmed down. Not to forget, September 11th 2001 has reminded us that we live in cities of flesh and stone. The ideas of their exoduses into virtual space has lasted just a semi decade. Moreover, it therefore has to be considered, that the Spring of 2000 has a significance in respect to the public and the academic discussion and assessment of the ICT potential. From today's perspective the date demands to take a break of visionary conceptions. These might have overstressed the utopian potential of ICT. Nevertheless the burst of the hype with all its economic consequences allows now to come back to a more differentiated consideration of ICT.

The new space of electronically mediated social interaction and communication has been and still is just a medial space created by an urban society. From the beginning of the discussion it therefore was asked, if and how the relationship between the urban society of today and the communicative potential of ICT could be formed in terms of politics. In the end, the crucial point has been less a question of technology but of how to create the new space socially. Do national differences in terms of individual needs and values should be reconsidered in order to get on the path? On the other hand, do particular situations on the city level surpass those differences? How important are the individual circumstances, needs and priorities of cities differing in size and economic power? How could the interface between the physical space and the virtual city should look like in order to manage the challenges of ICT in terms of its significant potential benefits and serious potential threats? Are there any patent remedies how ICT can be used in regard to ailing cities? It has been assumed that there is some kind of spatial interplay between ICT and urban development. This raises a whole range of items with respect to social issue of urban life. The information infrastructure must not only be managed and governed transparently, but it must be also available to all. To guarantee access this implies not only technical standards but also skill bases and in the end a design how the shift could be managed in a socially justifiable way. ICT needs social organisation.

At this policy point the Tele City Vision Project started in 1999 focussing on ICT perspectives of European urban actors such as decision makers and urban planners. Thus, TCV has not been a study about how many metres of fibre cable have been laid, it has not been a study about how many households now have access to the Internet. In the first line, TCV it has been a study about the use of ICT and its perception in the field of urban development policy. The overall objective was to examine and to compare on a European level and within different types of cities attitudes, behaviours and expectations of local urban actors in order to contribute to an improved understanding of how the potentials of ICTs are assessed in respect to their impact on the city.

TeleCity as Metaphor or the Project's Approach

It is a basis argument of the TeleCityVision project that ICT does not autonomously reshape society. Thus, the idea has not been to examine the impact of ICT on urban development in any direct way. Such an approach of technological determinism would be based on a linear notion that innovation leads to adoption of new technologies by society, and this, in turn, directly affects society. Instead, central to our survey stayed the focus on the awareness of administrative and political actors towards ICTs in different European cities. The approach of the project was to see technology as part or as an extension of society.

Social knowledge of individuals and the public social and political discourse shape the way IT are developed and applied within society.[2]

Let us explain the approach by the following example of the study. A direct impact on spatial issues has been rejected by the people asked in the first qualitative phase of our project. From a distance perspective, the outcomes could lead to the general impression that within the attitudes, expectations and hopes put into ICT, the fundamental problems of current urban development policies are reflected in a particular manner. What does this mean? Taking into account the estimates of the different groups the TCV project has interviewed, we observed different knowledge deficits regarding the potential of ICT. What is the reason for this? It has been ascertained in the qualitative phase that ICT initiatives are often due to the efforts of individuals. However, the quantitative survey demonstrated then that on the one hand the involved people from urban municipalities and city administrations were aware about the existence of an ICT strategy plan for their city, but on the other hand unable to name them precisely. What does this say? There are obviously public policy plans to guide information society developments, but the effectiveness of them is limited. The knowledge is very small. Out of 1560 respondents 433 answered that there is a formal ICT plan, nevertheless only 127 were able to name it. Drawing the conclusion that

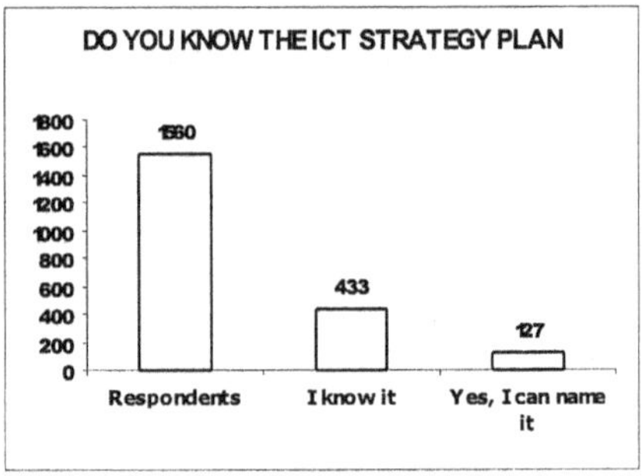

[2] In view of that the research work has been designed. There have been three main phases. The project's work started with a structural analysis of the current state of the urban system in Europe. The objective was to analyse and to evaluate regional and national development data in the seven participating countries of the project, Austria, France, Germany, Ireland, Netherlands, Norway, and Spain. The idea then was to select and classify cities in a tentative typology in preparation of the second work phase. For a qualitative case study three to four urban agglomerations per participating country were chosen. The plan was to capture different views and perceptions on ICT and the information age from different types of cities and to generate hypotheses. The results of the case studies were transferred into a questionnaire for the third phase, a quantitative survey, that should test the hypotheses.

this superficiality is based in individual ignorance might seem logical, but would be still a kind of rash judgement. It is too easy to claim that city administrators are not aware of the really important processes transforming the world. Thus, we have to try to understand, why actors are not prepared to inform themselves about ICT plans.

A perspective could be to see the interviewees as citizens of their cities. Professionals are not completely autonomous, but are part of their city as a whole. Both, the perception of ICT and the city is due to individual experiences and assessments. The personal technological belief in and the knowledge and private use of ICT should be seen as important determinants that influence the awareness towards the potentials of ICT. Moreover, as citizens, also urban actors bear in mind a very individual city conception, which determines their understanding and perception of urban politics. According to Kevin Lynch, one could point out that the urbanite is reading the city day-by-day (Lynch: 1960; 1985). The city conception combines the everyday picture of the city including mental maps everybody usually has and also the general idea of the city which is in conjunction with the social and economical surrounding. This includes the consciousness of physical structures and infrastructures, the awareness of economic conditions and functional characteristics, and the significance of size, hierarchy and location of the city within the urban system of Europe and throughout the world.

Nevertheless, the persons being asked by us about ICT and its possible impact on the city transfer their individual concept of city, including the professional knowledge about current urban problems etc., to the potential uses of ICT as well as to the question whether these potentials can be useful for the city. Such a transformation ideally implies that the individual concept of city flows into the understanding as well as the shaping of ICT. Apparently, we also see that numerous applications of ICT have a sort of urban structure. But how can such transformative act be understood in order to gain an understanding which goes beyond the construction of mono- causalities?

It has been noted that the way, we communicate about ICT, is accompanied by certain metaphors (Bolter 1996). Also the talk of the TeleCity has to be seen as such a metaphor which implies certain future expectations. Understood as little models of thought metaphors shape technical innovations. An example can been seen in the network metaphor. In regard to ICT it contains the idea to parallel numbers of PCs to a net. Moreover, the technical innovation vice versa provides new social models, for example the network society. From here, this migration of metaphors goes on and while there is a social shaping of technology, also the society is formed by the new technical innovations.

Taking the social shaping of technology approach seriously, it raises the question in respect to the TCV project, how the ascertained knowledge deficits of urban actors can be related to such transformative processes? One has to bear in mind, that a metaphorical use does not only shape the technological innovation but also the communication about the innovation itself. There is a steady translation process from one site to the other. It determines not only the everyday use and the individual perception of the medium as a symbol of a new era, but it also sets the frame for the shaping of guidelines and strategies in respect to the expected social potential of the technology.

The internet is for example clearly seen as the most prominent ICT tool. Meanwhile it has embodied numbers of different individual communication situations and styles which also create urban life. One could understand the internet as a medium which has established all kinds of new urban medial forms. It combines urban communication such as newspapers, postal services, radio, television, cinema etc. It provides urban services such as library or administrative services. Furthermore it contains different sorts of platforms which enable an individual communication of groups throughout the world. Its progressive communication density therefore has encouraged to compare it with an urban space. Moreover, it has been shown that there are two different metaphorical ways of talking about it referring to different understandings of its use. The image field of the so called data highway, implying the efficiency of regulated and purposeful transport, refers to an economic understanding of the internet while the field of navigation rather refers to a democratic understanding including values of free speech etc. (Bickenbach and Maye 1997).

Understanding the new virtual space of the internet as an medial enlargement there is obviously an analogy between the urban and the ICT discourse. Following the German sociologist Hartmut Häußermann (1993), many cities are today faced with the fundamental decision of whether they can still be understood as a collective social actor, as a space, which encloses all the inhabitants under the welfare of citizenship, or as a marketplace, where only those can participate who obey its laws, rules and behaviour patterns. The city is confronted with the a similar situation as the Tele City, whether it will be accessible to all, and thus a collective medium, or only to a few, who then can afford to control the access to the new urban enlargement. What does this analogy could imply for the understanding of the respondents' perceptions of ICT? Many of them named some ideas about ICT potentials in regard to the city. But still their assessments remained very unspecific. In the end might stay to question, whether urban actors have an useful city conception in the first line. If they do not, and this is most likely in respect to the crisis of the city, how shall they become aware of the ICT potentials for the city? Because of the complexity, city conceptions are mostly gained in parts due to what is currently seen as the most important issue. This implies that some urban structures and problems can be embodied and others must be left out. Therefore, the awareness towards ICT in regard to city related potentials is very selective, at least it includes whatever is seen as the most significant item according to the city conception of the individual respondent.

The question about the impact of ICT on the city is always a question what people bear in mind about the city itself. Thus, our two simple thesis in order to present the results of our study are:

- As long as city officials do not have an explicit city concept taking the urban structure and problems extensively as possible into account, there will be no adequate perception of the potential of ICT.

- Because ICT is a socially shaped technology of urban communication, the thinking about ICT and its potential linkages to the city is constantly demonstrating the vital situation European Cities are confronted with today.

TeleCity Rivalry

In the following presentation we would like to justify this by focussing on the issue of city competition.

As part of the consciousness of the urban stagnation, the increasing competition within cities as well as among different cities must be seen as a peremptory and particular problem field of the urban crisis. First of all, city competition again makes aware that national borders are losing their significance. Nevertheless, it would be an error to believe that the importance of cities and regions increases in proportion to the decrease of national regulatory frameworks. Rather, the increasing competition shows that the space among cities within the world-wide city network is becoming closer as more and more cities are finding themselves willingly or unwillingly forced to accept that challenge.

Under financial restrictions it is necessary today to use all existing resources in order to stimulate innovations and produce growth with the goal of establishing a significant position within the world-wide city network. In the past, rivalry among cities, was discussed mostly in regard to big cities. But now smaller cities are taking part in this rivalry, meaning that we must include them in our perspective. More and more small and medium-size cities are participating in the contest of obtaining a good „position within the international division of labour". More and more are competing in „functions of control and order" pertaining to adjacent regional areas and are taking part in the „competition for consumers and money" (Brandner 1997: 349). Not only "relevant European cities" such as Amsterdam and Vienna, but also other cities of Europe are in a struggle with each other. In our quantitative examination 78.1% of all respondents see the competition of the cities rising.

In line with this assumed increasing pressure of competition the visibility of each individual city within the international context becomes a vital issue. Visibility can be defined as the amount of attention a city is able to gain world wide, as for example, through the attraction of events and the number of visiting tourists (Franck 1998). While the Global Cities are obviously ever-present and draw more and more innovation and investment capital, national relevant cities like Leipzig or Oslo have to combat strongly the growing threat to their invisibility. Even more effort is needed by smaller cities. Interesting in this matter are the assessments within the quantitative survey of our project on the increase of importance of each actor's own city. Its contrariness gives a clearly insight on the powers that are combined by the awareness of an increased city competition.

At first glance a homogeneous picture is recognisable. The opinion of 74.1% in accumulated numbers of all people asked in cities over 100.000 inhabitants is that their city will gain in importance in the coming decade; in cities with less inhabitants this assumption increases slightly to 85.3% and in cities with a population under 50.000 inhabitants it increases somewhat again, to 86.2%. In regard to the future of their own city, people asked are undoubtedly optimistic. As our survey deals with city personnel, who probably identify themselves significantly with the city because of their professional work, this outcome seems plausible. However, one could ask further, for what reason does the optimism increase even more in smaller cities? Does a smaller size city allow another identification that furthers the optimism? Is there a better reading of the city possible? Or does the optimism of the smaller city mask its actual insignificance?

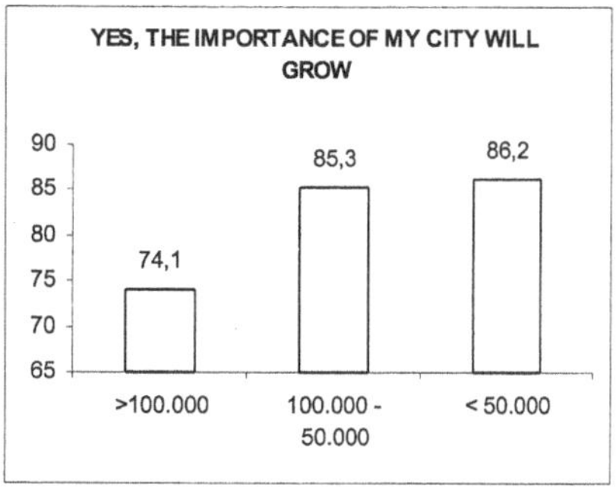

The optimistic picture can be seen in another perspective after further questioning. Comparing large and small cities, the assessment of changes in importance differs. In cities of over 100.000 inhabitants just 40.2 % of the people consider that the importance of smaller cities will increase, whereas 32.7 % think it is implausible that there will be any change and another 22.4 % even believe that there will be a decrease in the importance of small cities. This seems not only due to the arrogance of the city dweller. Similarly, the optimistic tone is reduced by the people from cities under 100.000 inhabitants. The tentative negative assessment of the city dwellers is shared regarding the importance of smaller cities. Observing cities under 50.000 the answers are again similar. Only 11% suppose that the importance of small cities will increase considerably, 35.1 % believe to some extent, 29.3% consider there will be no change, and 17.8% believe there will be a decline. Hence it follows that there are divergences between the confidence in regard to people's own city and the increase in importance of other cities. After all, every fifth person in cities under 100.000 believes that there will be a decrease in smaller cities, while, no matter on which level of city size, everyone assumes there will be a further increase of big cities. 70 % of all that were asked are convinced that the big city will gain further power. To summarise, in opposition to the mainstream assessment that all cities no matter what size, hierarchy or location they belong to, the urban employees of small cities believe on the hand in the future of their own city, but being asked about small cities in general, there is a mistrust in respect to their future importance.

Let us come to national differences. Because countries do not have equal numbers of big cities, it is clear that, from a comparative point of view, there are national differences. In those countries, such as Germany, with many cities of over 100.000 population, it is considered implausible that small cities will increase in importance. 38.5 % believe that the relevance of small cities will not change, and 27.4% even predict a decline of smaller towns in the ranking of cities. The Netherlands show a similar assessment as well. However, in both participating countries with the lowest density of urban population, Austria with only five cities over 100.000 inhabitants, and especially Norway with just four big cities, the assessments are rather reserved. The two countries with a more optimistic perspective are France, followed by Spain and than especially Ireland. 58.1% of all questioned from Ireland expect an increase of importance and another 29.7% even sees this growth as considerable in the coming years. Ireland has two cities of over 100.000 inhabitants, Dublin and Cork. Just 21% of

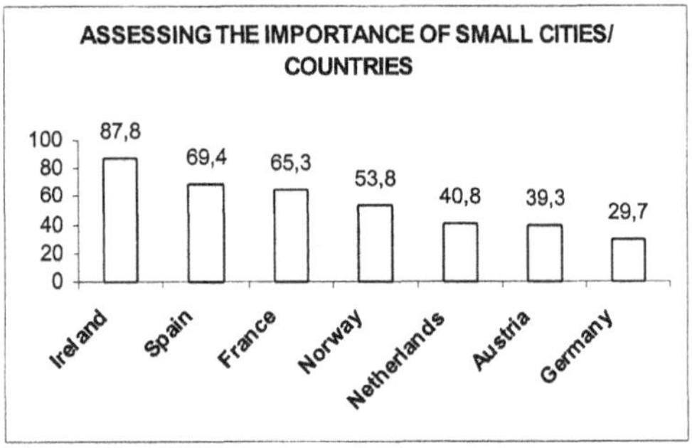

all who were asked came from one of these cities. Even in Spain, which has approximately as many big cities as France, one also believes in an improvement of small cities. What does this mean?

Understanding City Competition

All of these results underline that the competitive wind is felt stronger, just as the first optimistic self-assessment implied. In respect to ICT this also has been assessed positively. The Berlin Declaration on the Urban Future of 2000 stated quite optimistic and somehow technology driven: „Globalisation and the information technology revolution will increasingly create a borderless world with a new role for cities." and, further on: „Cities should embrace information and communication technologies and promote the lifelong education of all their citizens to become learning cities and to achieve global competitiveness."

But before we ask what role plays ICT in this contexts, the question remains, how competitive mentality works? The discrepancy between the *positive* self-assessment of small cities and the more sceptical view from outside reveals already the answer. It is possible to differentiate between an inner effect of city competition referring to the internal climate and an outer effect which is orientated to external appearance of the city.

The confidence in regard to the urban future of people's own city leads to the presumption that the competitive mentality not only hardens city life, but also affects positively the identification of citizens with their own city and its surroundings. For many peripheral cities in times where money is tight, the belief that something is changing is the initial prerequisite for a better future. This inner effect of city competition is the pre-condition for competition of cities among each other. People are moving closer together in order to be well prepared for what comes. Regarding the inner cohesion of cities, the principle works. It creates an urban identification in opposition to outer insecurity, thus enabling people's own city to better the climate within the city.

Vital in regard to the inner effect of the competitive principle are soft factors referring to the readability of city form. The surroundings in terms of neighbourhoods, offerings of leisure and culture, and also in light of the question whether the urban setting is worth living in, or suffers from pollution, can be affected by this part of the competitive principle (Grabow et al. 1995). Also, it implies the promotion of education and integration. Positive but also negative consequences in regard to political strategies can be the result of this impact. Cities suffering heavily under industrial wasteland will probably respond differently towards the competitive pressure than cities with a high standard of living and education. A negative effect of this side of the competition principle lies in the exclusion of the unknown, which is then identified with the insecurity of the space outside the city. But competition also provides the urge to become more homogeneous, motivates building walls around the city's own merits in order to be safe. In a dream of purity based on common sense, the internal effect can lead to expelling everyone who is not loyal to the rules of the city.

In the external direction the competitive principle develops a different dynamic. The goal is to better factors such as transportation infrastructure, employment structure and within this the creation of office space, the strategy aims to attract capital and therefore invests in external presentation of the cities for potential investors and tourists. Surely these factors have an effect on the city's interior as well. There are studies for example which maintain that the mechanisms of social control within a city are ruled more-and-more by tourists than by the original citizens. One can also study the effect by taking city guideline into account. There is a guideline in Tilburg in the Netherlands, a city of almost 200,000 inhabitants, which had presented itself at the beginning of our project as the "textile town". Nowadays, the guideline has changed, at least according to its web site: There it lies not only in a beautiful surrounding but in the heart of the booming region of Mid-Brabant. Both the inner and the outer effects of city competition cannot be considered separately. But it is still necessary to consider if the effects contradict each other.

One has to consider, if an increasing city competition is oriented exclusively to the outside, if one is only concerned about giving a signal to the world, one gambles with the threat to hide the problems behind an image of the city according to the standards of international representation. The understanding of how the city competition works gives us the hint that within the answers on the changing importance of actors' own cities there might be a certain gap between the perception of the internal structures of the city and the external presentation. In many cities, unbounding self-confidence unmasks itself sooner or later as a last scream for help in a city suffering financial catastrophe.

ICT and Urban Visions

In this danger of contradiction the results of the TeleCityVision project show that ICT plays a vital role. In regard to both the inner and the outer effect of competition however, the application of the new information and communication technologies offers potential. ICT certainly gives new impulses to business. But, the question remains, how these effects are assessed by the urban actors, in order to examine their perceptions of ICT? It might depend on the individual care of either the inner or outer effects of the city competition.

Especially internet presentations of cities and communities show the outer effect of city competition. Such presentations offer even economically weaker cities a possibility to present themselves, excluded from all inner conflicts and urban problems, as being a modern location. But also large stagnating industrial cities can refurbish their bad image as a successful city. The web site than becomes a multilingual visiting and business card of a municipality acting more-and-more as a private firm. The internet presence is part of an urban development policy, which acts as city marketing on the basis of public-private-partnerships, as many ICT-initiatives which have been studied during the project have proven. Thereby hints to big events are not allowed to be missed on the Web sites, as for example: "The famous Tilburg funfair is the largest in Europe!" They help to support the fragile identity of urbanites and also increase the acceptance of mistrustful investors from the outside. Besides the actual costs for festivals, expositions and other events, this allows, at least for a relative time period, to gain some attention on the international parquet for the urban competition. Today one talks more and more of a "festivalisation" of urban politics, which aligns its strategies only to the external presentation of the city (Häußermann et al.: 1993). As is being said, the reason for that is because there is no political advantage worth the effort taking classical urban issues into account. City conception is not related to experiences of city life, but to the outside. ICT is taking effect as a multiplier regarding this precarious situation.

As much as ICT promises equal chances for large and small cities to participate within the world-wide city network, general access due to the new technologies can also mean that city competition increases even more. What has been shown to be a capable answer in regard to the economical erosion of the cities, also even promotes this erosion. Therefore, it has already been proposed as a very discussible matter, to understand a loss in importance within the world-wide city network as a chance. It helps to concentrate on the reality of urban problems. However, the quantitative survey of our project shows that ICT will increase the struggle among cities according to the people asked. All-in-all 94.4 % believe that urban contests will be re-enforced by ICT, whereas 41.8% expect a high impact, 38.3% a medium, and 14.4% a rather low impact. Again, Ireland and Spain expect more rivalry caused by ICT. Moreover, in cities with more than 100.000 inhabitants one guesses that ICT will cause an increase. Around 45.6%, believe that the increase in competitive pressure will be significant. Compared to the assessments not taking ICT into account, a increase due to ICT can be ascertained.

The differences in the assessment demonstrate again the different effects of city competition principle. In regard to the outer presentation and its corresponding factors one sees the substantial potential of ICT. 88.6 % of all respondents take for granted that because of ICT more service business will settle within the cities, 60.2 % are also sure that ICT will certainly attract further industry. Otherwise, only 56.2%, see a possible impact caused by ICT in the question of attracting further residents to the cities, which can be related as a soft factor due to the inner effect. Besides the curiosity that especially France is optimistic in that point, also countries with a low population density as for example Spain, Ireland and especially Norway see at

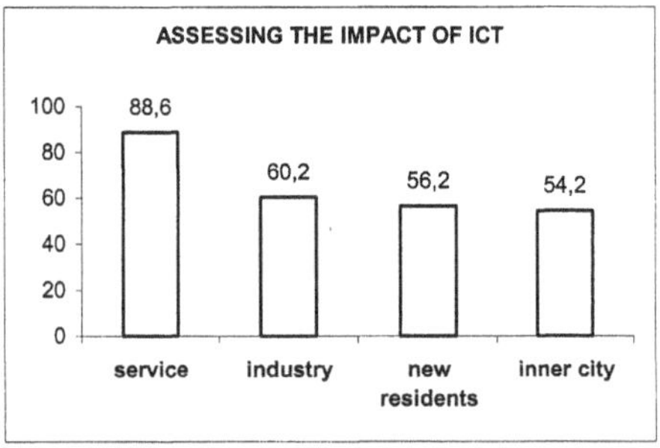

least a little potential on this issue. In Germany instead one does not see an activating factor in regard to residential growth. Furthermore just 54.2% of all people questioned suggest an impact of ICT in regard to the significance and shaping of the inner-city area. Although both aspects, the residential development and especially the formation of the inner city are very much ICT-related issues - the use of security technologies at public sites and streets exist today not only in the UK but in all of Europe– it is proven that the impact of ICT under the increasing competitive pressure relates substantially more to the external effects of the competition principle and its most relevant factors.

The qualitative investigation demonstrated that ICT in the minds of city planners and politicians at first meant attracting Big Business, before raising any questions regarding social structuring of the city for the purpose of improving inner-city climate and infrastructure. This is re-confirmed in the context of city ICT politics. The potential of ICT for attracting companies stands in the foreground, not city activity. 90.2% see ICT as a decisive instrument of city development politics for boosting economic development. And half of those of this opinion see ICT as a very decisive instrument. 84% are convinced that bringing companies to the city is the most important effect of an ICT engagement. The interviewees consider that ICT promises untold access possibilities to the world market. But it remains questionable how pressing inner infrastructure problems can be solved as long as they are only of secondary importance. However, it would be interesting to see how this assessment has changed since the year of 2000.

Two questions must be asked: how important are social issues in the context of ICT? Secondly, does ICT cause city planners to lose sight of spatial indicators when seeing the advantages of an improvement in city portrayal for the purpose of attracting companies? Of course, it will not suffice to reduce ICT to Internet sites. A city homepage will appear to reflect modernity and technical competence, that is, an improvement in internal communication structure, which in itself will be important for portraying the city. Quantitative questioning allows further differentiation. Beside the political-economic importance attached to ICT, people are convinced that ICT will also meaningfully affect city political communication processes. For example, in most countries there is a belief that the employment of ICT aims to make the political and administrative processes more transparent. In Norway, as the only country of the study, the respondents consider a higher effect of ICT on political transparency than on the improving of the economic situation. In opposition, the respondents from the Netherlands – the country compared to Norway in particular and to Europe in general with the highest population density - are the most reserved about this aspect.

But this basic assessment that urban policy will change does not allow evaluating in which direction this change will be. And this question, how much will city politics be changed by ICT, is the central point of city competition, because here outer and inner effects of the principle are obviously most visibly connected. The countries investigated - with the peculiar exception of the Netherlands again - are convinced that ICT will in general strengthen information flow and communication. 93.2% believe in this potential of ICT. On the other hand, it is assumed that social relationships will not be particularly enhanced through ICT, which is a subject in itself. Here only 19.6% of all respondents see an impact. But for city political processes it can be said that improvements in communication between city administration and citizens are expected. Again, especially the interviewees from Norway see a further chance, whereby it should be considered that the Scandinavian countries have a longer tradition in the democratic use of ICT, and that Norwegians in general know each other better than people in countries with a much larger population.

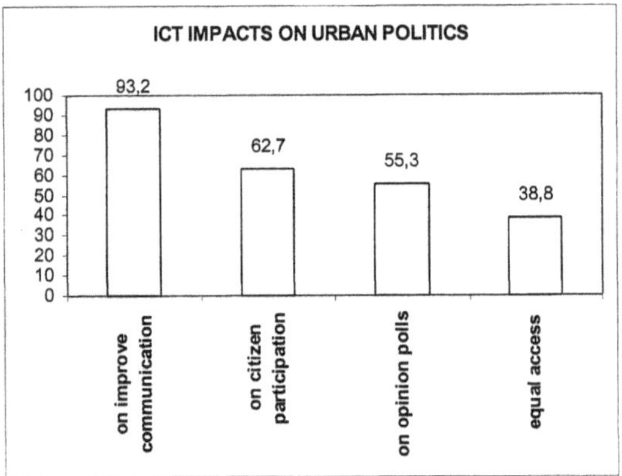

However, these positive expectations are subject to important limitations in two aspects. More communication in politics with the help of ICT is not the same as more influence. Even when the interviewees were convinced that there are more possibilities for citizens, politicians and administration officials to communicate among themselves, they reveal an uncertain confidence in regard to an increase of citizen participation in politics as a result of ICT.

Even though Spain and Germany are the usual opposites, only two-thirds or 62.7 % believe that ICT allows an increase in citizen participation. The results show that a certain scepticism is justified in spite of the positive assessment. Except for the confidence of the Spaniards, barely half think it probable that ICT will lead city administrators to take account of public opinion. Austria, Ireland and Norway have the same opinion here. The interviewees in France, Germany and the Netherlands are much more sceptical. The trust in growth of grass-roots democratic elements however must be questioned when the issue is the effect of ICT on access to education, work and services for all social classes of the city. Then things appear differently. Only a good third or 38.8 % still believe that ICT creates equal access opportunity for all social classes. The interviewees in France, Austria and Spain are still convinced, but this question is perceived in Germany, Ireland, the Netherlands and Norway with a certain suspicion.

However a change in the urban politics caused by ICT is seen. Its consequences remain rather vague. Once again, it must be asserted that ICT is still understood as a one-way-street citizens as well as administrators can use in order to inform themselves about each other. But we must realise that political influence by citizen participation cannot be shown to be a potential of ICT, because all citizens do not have equal access to information through ICT. According to the perspective of the respondents the use of information seems to be restricted to the economic aspects. It enforces the external orientation of urban politics, and it must be doubted that all social groups will have a profit out of it. The majority of the respondents even believes that ICT will increase the gap between the rich and the poor. In numbers are this 57.8%. What do we learn from this? The scope and depth of ICT perceptions by urban administrative and political actors served within the idea of the TeleCityVision proj-

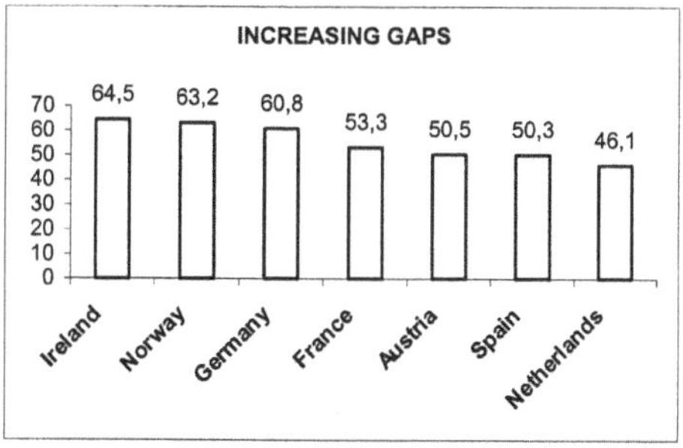

ect as indicators for the preparedness to contribute to the transformation process.

Conclusion

To be aware of newly emerging and rapidly changing technologies and to adopt them to enhance the everyday life of urbanites, obviously means to be in the midst of a global communication process where the individual or collective perception ability of what is going on in the world of technology and science, the public culture, defines the bandwidth of participation within this process. The global communication process represents the space where one talks about and discusses the technological revolution and where personal experiences are exchanged. Does not to participate means to neglect an important possibility to get new and fresh ideas for improving urban life? The use of ICT could be seen as a medial enlargement of the city beyond its material and already existing informational resources and borders in order to handle and to create the complex urban relationships and communication processes which always have been the wellsprings of urban social, scientific, technological and economic innovation. However, time has shown that this medial geography is hardly able to solve social injustices or urban problems like there always have been.

The perception of the urban actors on the impacts of ICTs have been rather sober than enthusiastic. In our reading of the results of TeleCityVision study the increased city competition is clearly seen effected by the transformation to a global information society. But on the other hand ICT was hardly seen as a panacea to solve urban stagnation as it has been promoted in the middle of the 1990s in the political and the media discourse. Even if there have been knowledge deficits regarding the potential of ICT, the assessments show within the framework of an increasing city competition particularly that besides the constraint of being orientated towards the outside, urban actors are still much aware of the inner structure and of more local urban problems of their cities. Planning itself in regard to housing or transportation will, in the opinion of the interviewees, continue to take place "in the heads" and cannot be replaced by computer generated procedures. Actors know about the complexity of the urban crisis within their individual conception of city, but they are also aware of their limited potentialities in regard to empty finance resources.

Thus, discrepancies between the awareness of technological potentials might rather find their explanations by the high expectations of visionary concepts and the limitation of policy capabilities. Over the last decades the political understanding of urban space has changed. In reality characteristics of the Keynesian Welfare State are replaced by 'entrepreneuralism' as the main design of urban action. A trend towards the post-Fordist city pretending to be a playful space of individuality but being beneath its surface a space of control and social and spatial separation (Christopherson 1994: 409) raises today the question how far these developments of creating feel good atmospheres can be related to the rise of ICT? The logic of contemporary place marketing as a critical element to derive competitive advantage in the international rivalry of cities for investment meets the virtual space as it helps to produce a certain symbolic purity that emphasises the dissociation of depression and industrial decline. What does this mean? The overestimation of ICT contains the threat to suppress and forget the reality of city problems. Today we witness, urgent individual social problems are more and more isolated from the realm of urban politics. The managing of the space within the post-Fordist city focuses instead on the separation of different kinds of people in order to create spaces of consumption for tourists and consumers. In this respect ICT turns out to be not a solution, but a tool of social sorting and thus for the displacement of urban problems. But in the end, the recognition of city-related potential of ICT depends on the conception of the individual city urban politics have in mind for the motifs of their actions. They have to decide if and how different economical interests within the contemporary city can be balanced, for example between potentially consumers and non-consumers or in spatial terms between places of consumption and those publicly accessible to all. As a matter of fact, such a decision could turn out as the most urgent attempt to guarantee the social peace for the city of tomorrow.

References

Bangemann, Martin (1997): “A New World Order for Global Communications”, The Need for an International Charter, http://www.ispo.cec.be/infosoc/promo/speech/geneva.html

“Berlin Declaration on the Urban Future”, Declaration of the official (governmental) URBAN 21 Conference 6 July 2000, http://www.laneta.apc.org/hic-al/cartaseng_3.html

Brandner, Birgit (1997), „Städtekonkurrenz, Stadtprofilierung und Kultur“. Europäische Perspektiven, in: Kommunikationswelten: wissenschaftliche Perspektiven zur Medien- und Informationsgesellschaft, Innsbruck, Wien, pp. 347-375.

Bickenbach, Matthias and Harun Maye 1997: „Zwischen fest und flüssig.“ Das Medium Internet und die Entdeckung seiner Metaphern, in: Gräf, Lorenz (eds.): Soziologie des Internet: Handeln im elektronischen Web-Werk. Frankfurt/Main; New York : Campus Verlag, pp. 80-98

Castells, Manuel (1989). “The Informational City”: Information Technology, Economic Restructuring, and the Urban-Regional Process. Oxford: Blackwell

Castells, Manuel. (2001). “Das Informationszeitalter: Wirtschaft, Gesellschaft, Kultur“, Teil 1: Der Aufstieg der Netzwerkgesellschaft. Opladen: Leske und Budrich.

Christopherson, S. (1994). The fortress city. Privatized spaces, consumer citizenship. Post-Fordism. A reader. A. Amin. Oxford, Blackwell: 409-427.

Franck, Georg (1998): “Ökonomie der Aufmerksamkeit”, München: Hanser

Gandy, O. (1989). „The surveillance society. Information, technology and bureaucratic social control”, in: Journal of Communication 39(3): pp. 61-76

Häußermann, Hartmut; Siebel, Walter (Hrsg.) 1993: „Festivalisierung der Stadtpolitik. Stadtentwicklung durch große Projekte“, in: Leviathan - Sonderheft: 13. Opladen: Westdeutscher Verlag

Melody, William H. (1996): “Toward a framework for designing information society policies”, in: Telecommunication Policy, Vol. 20, No. 4, pp. 243-259

Leggewie, Claus; Christa Maar (1998): „Internet & Politik“, Von der Zuschauer zur Beteiligungsdemokratie, Köln: Bollmann

Lynch, Kevin (1960): “The image of the city”, Cambridge, Mass.: MIT Pr.

Lynch, Kevin (1985): “Good city form”. Cambridge, Mass.: MIT Pr.

Saskia Sassen (2000): „Machtbeben“, Wohin führt die Globalisierung, Stuttgart München: Deutsche Verlags-Anstalt

Contributors

Galit Cohen-Blankshtain, Ph.D., graduated in Geography and environmental planning at the Hebrew University of Jerusalem, Israel, and received her PhD at the Department of Spatial Economics at the Free University, Amsterdam. Her PhD dissertation title was: "Urban Futures and ICT policies: Perceptions of Urban Decision Makers". She is interested in various aspects of public policy issues and policymaking process. More specifically, her main research interests are urban ICT policies, transport policies, modeling transport behavior, and urban governance. Currently she is visiting at Alfred Taubman Center for State and Local Government, Kennedy School of Government, Harvard University.

Hans-Liudger Dienel, Dr., Dipl.-Ing., was born 1961. He is Acting Director of the Zentrum Technik und Gesellschaft of the Technische Universität Berlin since November 1995. After his degrees in History, Philosophy and Mechanical Engineering (M.A. 1988, Dipl.-Ing. 1989), he worked as an Assistant Professor at the Technical University Munich (1990, 1992), Institute for the History of Technology and passed his Ph.D. in 1993. From 1993 to 1995, Dienel headed the transport research group in the research institute of the leading Museum for Science, Technology and Industry in Germany (Deutsches Museum) in Munich. He is currently working on models of urban life, urban transport and participation theories.

Bernd Fischer, Dipl.-Pol., born in 1967, studied Political Sciences, Law, Sociology and Geography at University of Frankfurt/M. He worked as journalist. Since 1996 he is research fellow at Berlin Institute for Social Research. He was lecturer in Sociology at Alice-Salomon-University for Applied Science, Berlin and is lecturer at various vocational colleges for health care services in Berlin. Bernd Fischer organised and co-ordinated the TeleCityVision project together with Frank Helten. He has published on social changes in the information society especially in private households, health care services, technological innovations in organisations and urban development. Currently he is working on a project about video surveillance in the publicly accessible space (Urbaneye).

Stephen Graham is Professor of Urban Technology at Newcastle University's School of Architecture, Planning and Landscape (SAPL) in the UK. Since 2002 he has been Deputy Director of the Global Urban Research Unit (GURU). His research develops a critical and 'socio-technical' perspectives to the reconfiguration of urban infrastructures and mobility systems; the growth of urban surveillance; the technological dimensions of social exclusion; the development of urban technology strategies; the 'relational' turn in urban theory; and the interconnections between war, terrorism and cities. He is the co-author of Telecommunications and the City (1996), Splintering Urbanism (2001) (both with Simon Marvin), editor of the Cybercities Reader (2003), and co-editor of Managing Cities (1995) and a forthcoming Blackwell book Cities, War and Terrorism (2004).

Frank Helten, Dr., Dipl.rer.soc., born in 1942, studied Economics, Engineering and Social Sciences at the Technical University of Berlin and at Ruhr-Universität Bochum. He has worked on urban problems since 1971 at the Urban Planning Office of Prof. Machtemes in Düsseldorf and at COMEDOR in Algiers (1973-1975). Lecturer and research fellow (urban and regional sociology) at Freie Universität Berlin and at Gesamthochschule/Universität Kassel (1976-1980). Research fellow at Technische Universität Berlin (sociology of technology, sociology of work and organisation, urban sociology, social research methodology (1980-1985). Member and research fellow (1986-1992): Interdisciplinary Research Group (Deutsche Forschungsgemeinschaft) "Konstruktionshandeln" at Technische Universität Berlin. 1993-1996: Project Manager at LIPS (Institute for Social Research), project-co-ordinator LVA-Sachsen. Currently he is directing 'mt helten research' and is senior research fellow and project-coordinator at BIS. Publications in the fields of urban sociology, sociology of technology, organisation and communication.

Leon Hempel, MA, studied German and Comparative Literature and Political Science in Berlin. After his graduation 1999 he started to work at the Centre Technology and Society. As a research assistant he worked for the TeleCityVision project on the impact of information and communication technologies on urban changes. Since September 2001 he is research assistant in the project Urbaneye on video surveillance in public accessible space that is co-ordinated by the Centre.

Santiago Lorente, Ph.D., Prof., Master of Arts in Sociology, Loyola University, Chicago, USA, 1966; Doctorate in Sociology, Universidad Complutense, Madrid, Spain, 1983; Director of the Consulting Firm EDIS, Madrid, 1970-1975; Director of the Audience Research Center, Spanish Television, 1981; Full professor of Sociology of ICTs at the Telecommunication School, Universidad Politécnica de Madrid, 1977 up to date. Has published 11 books or chapters of books, and over a hundred articles in subjects related to the impact of ICTs in society. Regarding the house, he is an expert on Domotics from a sociological point of view, having published the book "La Casa Inteligente" (The intelligent house) in 1991, is a member of the Spanish Committee on Domotics. Presently he is preparing two books: one on the intelligent home, from the sociological perspective, and another on the impact of ICTs on the family institution in the context of the information society.

Simon Marvin, Prof., is co-director of SURF and the United Utilities Chair of Sustainable Urban and Regional Development at the University of Salford. Marvin is an acknowledged national and international expert on the changing relations between cities and regions and infrastructure networks in a period of rapid technological change, environmental concern and institutional restructuring. He has published widely on cities, regions and technological change. Recent books include "Guy, S., Marvin, S.J,. and Moss. T. (2001) Urban Infrastructure in Transition: Networks, Buildings and Plans" and "Graham, S.D.N and Marvin, S.J, (2001) Splintering Urbanism: Networked Infrastructures, Technological Mobilities and the Urban Condition". He has led on a number of projects examining approaches to understanding urban and regional futures and the challenges they provide for future forms of governance. Simon Marvin recently completed a residency within the Urban Policy Unit of ODPM as part of a Treasury Placement Scheme during which time he examined the way in which a variety of central departments understand and conceptualise cities and city-regions and the implications for national approaches to sub-national policy and governance issues.

André van der Meer, Dr., (1953) graduated in Political Science in 1983 and obtained a Master's degree in urban management in 2000. He is a senior researcher for ICT and Urban Development at the European Institute for Comparative Urban Research (Euricur) at the Economic Faculty of the Erasmus University of Rotterdam. Two of the main research activities at the time are: "MUTEIS: macro-economic and urban trends in the European information society" and "E-Government: a study into the information society strategies of eight cities in Europe and South Africa". Previously, he has been working for the University of Amsterdam and the Dutch national government. In the 1990s he been the chief information officer (CIO) of the city of The Hague, where he now holds a management position for ICT and European relations. André van der Meer took a leading role in the take up of new technologies by local authorities in the European Union. He was one of the founders of Telecities (a co-operative network of cities) and served as: Chairman of the Major Cities of Europe IT Users group (1993-96), President of Telecities (1996-98), Chairman of the Infocities management board (1996-1998), Chairman of the European Digital Cities management board (1997-1999), Chairman of the Eurocities

Technology Policy Forum (1998-2001). He has been involved in several European projects, such as Equality, Infocities, EDC (European Digital Cities partnership), VirtuHalls, e-CT (electronic calls for tenders).

Peter Nijkamp is professor in regional and urban economics and in economic geography at the Free University, Amsterdam. His main research interests cover plan evaluation, multicriteria analysis, regional and urban planning, transport systems analysis, mathematical modelling, technological innovation, and resource management. In the past years he has focused his research in particular on quantitative methods for policy analysis, as well as on behavioural analysis of economic agents. He has a broad expertise in the area of public policy, services planning, infrastructure management and environmental protection. In all these fields he has published many books and numerous articles. He is member of editorial boards of more than 20 journals. He has been visiting professor in many universities all over the world. He is past president of the European Regional Science Association and of the Regional Science Association International. He is also fellow of the Royal Netherlands Academy of Sciences, and is immediate past vice-president of this organisation. Since June 2002 he serves as president of the governing board of the Netherlands Research Council (NWO).

Florian Rötzer, born in 1953, he is a theorist and art critic who lives in Munich. He published several books on electronic art and the digital era. He prepared five special issues of the cultural journal Kunstforum and was editor of the communications magazine "Tumult." He is also a curator of media exhibitions and events, among them "Digitale Schein" in Munich, 1991 and "Telepolis" in Luxemburg 95. Among other publications he is author of the book “Telepolis - Urbanity in the digital age” and editor of several edited books on this subject. Currently he is editor of the electronic journal "Telepolis". His resent publications are about the interplay of media and violence.

Saskia Sassen is the Ralph Lewis Professor of Sociology at the University of Chicago, and Centennial Visiting Professor at the London School of Economics. She is currently completing her forthcoming book "Denationalization : Territory, Authority and Rights in a Global Digital Age" based on her five year project on governance and accountability in a global economy. She has also just completed for UNESCO a five-year project on sustainable human settlement for which she set up a network of researchers and activists in over 50 countries. Her most recent books are "Guests and Aliens" and the edited "Global Networks, Linked Cities". "The Global City" is out in a new fully updated edition in 2001. Her books are translated into fourteen languages. She serves on several editorial boards and is an advisor to several international bodies. She is a Member of the Council on Foreign Relations, and Chair of the new Information Technology, International Cooperation and Global Security Committee of the Social Science Research Council (USA).

Manuel Valenzuela, Prof. Dr. Lead researcher and manager of the Research Laboratory on Urban and Social Geography, got his PhD degree at the Complutense University, Madrid (1973) and was appointed as professor of Human Geography at the Autonomous University of Madrid in 1981. From the beginning of his academic career he has developed two research lines: Urban Geography and Geography of Tourism and Leisure. He belongs to the advisory boards of scientific geographic reviews at the Spanish level and at the European one and performs as referee for public research agencies and scientific journals. He is a member of the committee for the International Geography Award "Vautrin Lud". Prof. Valenzuela has written or coordinated, alone or in collaboration, twelve books, forty books chapters, fifty journals papers and a similar number of contributions to proceedings on topics such as urban environment, urban planning, housing, leisure, urban peripheries and rural-urban fringes. He is member of several networks on urban problems at European (ENHR, EURA, UNICA) and worlwide levels (UGI). He belongs to the Directory Board of the Real Sociedad Geográfica and to the Spanish Committee of the International Geographical Union (U.G.I.). Prof. Valenzuela has been for sixteen years a member of international organizations such as the

UNESCO Man and Biosphere Program and an active member of the Spanish Committee of the United Nations Habitat Programme since 1996.

Carmen Vázquez, Ph.D. Associate prof., Licenciate in Geography, Autonomous University of Madrid, 1987, Doctorate in Geography, Autonomous University of Madrid, 1996. Part-time associate professor of Geography at the Autonomous University of Madrid (1995-1996), at the Faculty of Humanities, San Pablo-CEU University, Madrid (1996-1999), and at the Faculty of Humanities, Carlos III University, Madrid (1998-1999). Full time associate professor of Geography at the Faculty of Humanities, University of Castilla-La Mancha, Cuenca, Spain, 1999 up to date. She started her academic career working on Urban Social Geography topics dealing with urban social segregation problems and, later on, she broadened her scope towards Transport Infrastructure and Information and Communication Technologies (ICTs) related with cities. She has published 1 book and 11 chapters of books, and around eighteen articles in both Spanish and European journals and a similar number of contributions to proceedings. Since the last six years she has devoted her attention to the impact of ICTs on urban settlements.

Willem van Winden is in the Erasmus University Rotterdam, Faculty of Economics, department of Regional & Transport Economics. He has been involved in numerous comparative economic development studies, market studies, and economic impact studies. He developed cluster strategies for a number of large European cities. He is specialised in local/urban technology strategies and policies. He studied technology strategies for a number of European and South African cities, and currently participates in EC projects and other international research projects. He is (co) author of several books and a number of scientific articles in the field of ICT and urban development.